A Research Strategy for Environmental, Health, and Safety Aspects of **ENGINEERED NANOMATERIALS**

Committee to Develop a Research Strategy for Environmental, Health, and Safety Aspects of Engineered Nanomaterials

Board on Environmental Studies and Toxicology

Board on Chemical Sciences and Technology

Division on Earth and Life Studies

National Materials and Manufacturing Board

Division on Engineering and Physical Sciences

NATIONAL RESEARCH COUNCIL
OF THE NATIONAL ACADEMIES

THE NATIONAL ACADEMIES PRESS
Washington, D.C.
www.nap.edu

THE NATIONAL ACADEMIES PRESS 500 Fifth Street, NW Washington, DC 20001

NOTICE: The project that is the subject of this report was approved by the Governing Board of the National Research Council, whose members are drawn from the councils of the National Academy of Sciences, the National Academy of Engineering, and the Institute of Medicine. The members of the committee responsible for the report were chosen for their special competences and with regard for appropriate balance.

This project was supported by Contract EP-C-09-003 between the National Academy of Sciences and the U.S. Environmental Protection Agency. Any opinions, findings, conclusions, or recommendations expressed in this publication are those of the authors and do not necessarily reflect the view of the organizations or agencies that provided support for this project.

International Standard Book Number-13: 978-0-309-25328-4
International Standard Book Number-10: 0-309-25328-4

Additional copies of this report are available for sale from the National Academies Press, 500 Fifth Street, NW, Keck 360, Washington, DC 20001; (800) 624-6242 or (202) 334-3313; http://www.nap.edu/.

Printed in the United States of America

THE NATIONAL ACADEMIES

Advisers to the Nation on Science, Engineering, and Medicine

The **National Academy of Sciences** is a private, nonprofit, self-perpetuating society of distinguished scholars engaged in scientific and engineering research, dedicated to the furtherance of science and technology and to their use for the general welfare. Upon the authority of the charter granted to it by the Congress in 1863, the Academy has a mandate that requires it to advise the federal government on scientific and technical matters. Dr. Ralph J. Cicerone is president of the National Academy of Sciences.

The **National Academy of Engineering** was established in 1964, under the charter of the National Academy of Sciences, as a parallel organization of outstanding engineers. It is autonomous in its administration and in the selection of its members, sharing with the National Academy of Sciences the responsibility for advising the federal government. The National Academy of Engineering also sponsors engineering programs aimed at meeting national needs, encourages education and research, and recognizes the superior achievements of engineers. Dr. Charles M. Vest is president of the National Academy of Engineering.

The **Institute of Medicine** was established in 1970 by the National Academy of Sciences to secure the services of eminent members of appropriate professions in the examination of policy matters pertaining to the health of the public. The Institute acts under the responsibility given to the National Academy of Sciences by its congressional charter to be an adviser to the federal government and, upon its own initiative, to identify issues of medical care, research, and education. Dr. Harvey V. Fineberg is president of the Institute of Medicine.

The **National Research Council** was organized by the National Academy of Sciences in 1916 to associate the broad community of science and technology with the Academy's purposes of furthering knowledge and advising the federal government. Functioning in accordance with general policies determined by the Academy, the Council has become the principal operating agency of both the National Academy of Sciences and the National Academy of Engineering in providing services to the government, the public, and the scientific and engineering communities. The Council is administered jointly by both Academies and the Institute of Medicine. Dr. Ralph J. Cicerone and Dr. Charles M. Vest are chair and vice chair, respectively, of the National Research Council.

www.national-academies.org

Committee to Develop a Research Strategy for Environmental, Health, and Safety Aspects of Engineered Nanomaterials

Members

Jonathan M. Samet (*Chair*), University of Southern California, Los Angeles
Tina Bahadori, American Chemistry Council, Washington, DC (until May 2012)
Jurron Bradley, BASF, Florham Park, NJ
Seth Coe-Sullivan, QD Vision, Inc., Lexington, MA
Vicki L. Colvin, Rice University, Houston, TX
Edward D. Crandall, University of Southern California, Los Angeles
Richard A. Denison, Environmental Defense Fund, Washington, DC
William H. Farland, Colorado State University, Fort Collins
Martin Fritts, SAIC-Frederick, Frederick, MD
Philip Hopke, Clarkson University, Potsdam, NY
James E. hutchison, University of Oregon, Eugene
Rebecca D. Klaper, University of Wisconsin, Milwaukee
Gregory V. Lowry, Carnegie Mellon University, Pittsburgh, PA
Andrew Maynard, University of Michigan School of Public Health, Ann Arbor
Gunter Oberdorster, University of Rochester School of Medicine and Dentistry, Rochester, NY
Kathleen M. Rest, Union of Concerned Scientists, Cambridge, MA
Mark J. Utell, University of Rochester School of Medicine and Dentistry, Rochester, NY
David B. Warheit, DuPont Haskell Global Centers for Health and Environmental Sciences, Newark, DE
Mark R. Wiesner, Duke University, Durham, NC

Staff

Eileen Abt, Project Director
Tina Masciangioli, Senior Program Officer
Erik Svedberg, Senior Program Officer
Keegan Sawyer, Associate Program Officer (until August 2011)
Keri Schaffer, Research Associate
Norman Grossblatt, Senior Editor
Mirsada karalic-loncarevic, Manager, Technical Information Center
Radiah Rose, Manager, Editorial Projects
Tamara Dawson, Program Associate
Orin Luke, Senior Program Assistant (until June 2011)

Sponsor

U.S. Environmental Protection Agency

[1]This study was planned, overseen, and supported by the Board on Environmental Studies and Toxicology.

OTHER REPORTS OF THE BOARD ON ENVIRONMENTAL STUDIES AND TOXICOLOGY

Macondo Well–Deepwater Horizon Blowout: Lessons for Improving Offshore Drilling Safety (2012)
Feasibility of Using Mycoherbicides for Controlling Illicit Drug Crops (2011)
Improving Health in the United States: The Role of Health Impact Assessment (2011)
A Risk-Characterization Framework for Decision-Making at the Food and Drug Administration (2011)
Review of the Environmental Protection Agency's Draft IRIS Assessment of Formaldehyde (2011)
Toxicity-Pathway-Based Risk Assessment: Preparing for Paradigm Change (2010)
The Use of Title 42 Authority at the U.S. Environmental Protection Agency (2010)
Review of the Environmental Protection Agency's Draft IRIS Assessment of Tetrachloroethylene (2010)
Hidden Costs of Energy: Unpriced Consequences of Energy Production and Use (2009)
Contaminated Water Supplies at Camp Lejeune—Assessing Potential Health Effects (2009)
Review of the Federal Strategy for Nanotechnology-Related Environmental, Health, and Safety Research (2009)
Science and Decisions: Advancing Risk Assessment (2009)
Phthalates and Cumulative Risk Assessment: The Tasks Ahead (2008)
Estimating Mortality Risk Reduction and Economic Benefits from Controlling Ozone Air Pollution (2008)
Respiratory Diseases Research at NIOSH (2008)
Evaluating Research Efficiency in the U.S. Environmental Protection Agency (2008)
Hydrology, Ecology, and Fishes of the Klamath River Basin (2008)
Applications of Toxicogenomic Technologies to Predictive Toxicology and Risk Assessment (2007)
Models in Environmental Regulatory Decision Making (2007)
Toxicity Testing in the Twenty-first Century: A Vision and a Strategy (2007)
Sediment Dredging at Superfund Megasites: Assessing the Effectiveness (2007)
Environmental Impacts of Wind-Energy Projects (2007)
Scientific Review of the Proposed Risk Assessment Bulletin from the Office of Management and Budget (2007)
Assessing the Human Health Risks of Trichloroethylene: Key Scientific Issues (2006)
New Source Review for Stationary Sources of Air Pollution (2006)
Human Biomonitoring for Environmental Chemicals (2006)
Health Risks from Dioxin and Related Compounds: Evaluation of the EPA Reassessment (2006)
Fluoride in Drinking Water: A Scientific Review of EPA's Standards (2006)
State and Federal Standards for Mobile-Source Emissions (2006)

Superfund and Mining Megasites—Lessons from the Coeur d'Alene River Basin (2005)
Health Implications of Perchlorate Ingestion (2005)
Air Quality Management in the United States (2004)
Endangered and Threatened Species of the Platte River (2004)
Atlantic Salmon in Maine (2004)
Endangered and Threatened Fishes in the Klamath River Basin (2004)
Cumulative Environmental Effects of Alaska North Slope Oil and Gas Development (2003)
Estimating the Public Health Benefits of Proposed Air Pollution Regulations (2002)
Biosolids Applied to Land: Advancing Standards and Practices (2002)
The Airliner Cabin Environment and Health of Passengers and Crew (2002)
Arsenic in Drinking Water: 2001 Update (2001)
Evaluating Vehicle Emissions Inspection and Maintenance Programs (2001)
Compensating for Wetland Losses Under the Clean Water Act (2001)
A Risk-Management Strategy for PCB-Contaminated Sediments (2001)
Acute Exposure Guideline Levels for Selected Airborne Chemicals (twelve volumes, 2000-2012)
Toxicological Effects of Methylmercury (2000)
Strengthening Science at the U.S. Environmental Protection Agency (2000)
Scientific Frontiers in Developmental Toxicology and Risk Assessment (2000)
Ecological Indicators for the Nation (2000)
Waste Incineration and Public Health (2000)
Hormonally Active Agents in the Environment (1999)
Research Priorities for Airborne Particulate Matter (four volumes, 1998-2004)
The National Research Council's Committee on Toxicology: The First 50 Years (1997)
Carcinogens and Anticarcinogens in the Human Diet (1996)
Upstream: Salmon and Society in the Pacific Northwest (1996)
Science and the Endangered Species Act (1995)
Wetlands: Characteristics and Boundaries (1995)
Biologic Markers (five volumes, 1989-1995)
Science and Judgment in Risk Assessment (1994)
Pesticides in the Diets of Infants and Children (1993)
Dolphins and the Tuna Industry (1992)
Science and the National Parks (1992)
Human Exposure Assessment for Airborne Pollutants (1991)
Rethinking the Ozone Problem in Urban and Regional Air Pollution (1991)
Decline of the Sea Turtles (1990)

Copies of these reports may be ordered from the National Academies Press
(800) 624-6242 or (202) 334-3313
www.nap.edu

BOARD ON CHEMICAL SCIENCES AND TECHNOLOGY

Other Reports of the Board on Chemical Sciences and Technology

Prudent Practices in the Laboratory: Handling and Management of Chemical Hazards, Revised Edition (2011)
Promoting Chemical Laboratory Safety and Security in Developing Countries (2010)
Research at the Intersection of the Physical and Life Sciences (2010)
BioWatch and Public Health Surveillance: Evaluating Systems for the Early Detection of Biological Threats: Abbreviated Version (2010)
Strengthening High School Chemistry Education Through Teacher Outreach Programs: A Workshop Summary to the Chemical Sciences Roundtable (2009)
Catalysis for Energy: Fundamental Science and Long-Term Impacts of the U.S. Department of Energy Basic Energy Science Catalysis Science Program (2009)
Effectiveness of National Biosurveillance Systems: BioWatch and the Public Health System: Interim Report (2009)
A Framework for Assessing the Health Hazard Posed by Bioaerosols (2008)
Disrupting Improvised Explosive Device Terror Campaigns: Basic Research Opportunities: A Workshop Report (2008)
Test and Evaluation of Biological Standoff Detection Systems: Abbreviated Version (2008)

National Materials and Manufacturing Board

Members

Robert H. Latiff (*Chair*), R. Latiff Associates, Alexandria, VA
Denise F. Swink (*Vice-Chair*), Independent Consultant, Germantown, MD
Peter R. Bridenbaugh, NAE, Retired, ALCOA, Boca Raton, FL
Valerie M. Browning, ValTech Solutions, LLC, Port Tobacco, MD
Yet-Ming Chiang, NAE, Massachusetts Institute of Technology, Cambridge, MA
Paul Citron, NAE, Retired, Medtronic, Inc., Minnetonka, MN
George T. (Rusty) Gray, II, Los Alamos National Laboratories, Los Alamos, NM
Carol A. Handwerker, Purdue University, West Lafayette, IN
Thomas S. Hartwick, Independent Consultant, Snohomish, WA
Sundaresan Jayaraman, Georgia Institute of Technology, Atlanta, GA
David W. Johnson, Jr., NAE, Stevens Institute of Technology, Bedminster, NJ
Thomas King, Oak Ridge National Laboratories, Oak Ridge, TN
Michael F. McGrath, Analytic Services Inc., Arlington, VA
Nabil Nasr, Golisano Institute for Sustainability, Rochester, NY
Paul S. Peercy, NAE, University of Wisconsin-Madison, Madison, WI
Robert C. Pfahl, Jr., International Electronics Manufacturing Initiative, Herndon, VA
Vincent J. Russo, Aerospace Technologies Associates, LLC, Dayton, OH
Robert E. Schafrik, GE Aviation, Cincinnati, OH
Kenneth H. Sandhage, Georgia Institute of Technology, Atlanta, GA
Haydn Wadley, University of Virginia, Charlottesville, VA
Steven Wax, Independent Consultant, Reston, VA

Staff

Dennis Chamot, Acting Director
Erik B. Svedberg, Senior Program Officer
Ricky D. Washington, Executive Assistant
heather lozowski, Financial Associate
Laura Toth, Program Assistant

OTHER REPORTS OF THE NATIONAL MATERIALS AND MANUFACTURING BOARD

Opportunities in Protection Materials Science and Technology for Future Army Applications (2011)
Materials Needs and R&D Strategy for Future Military Aerospace Propulsion Systems (2011)
Research Opportunities in Corrosion Science and Engineering (2010)
Assessment of Corrosion Education (2009)
Proceedings of a Workshop on Materials State Awareness (2008)
Integrated Computational Materials Engineering: A Transformational Discipline for Improved Competitiveness and National Security (2008)
Managing Materials for a Twenty-first Century Military (2008)
A Path to the Next Generation of U.S. Bank Notes: Keeping Them Real (2007)
Assessment of Millimeter-Wave and Terahertz Technology for Detection and Identification of Concealed Explosives and Weapons (2007)
Fusion of Security System Data to Improve Airport Security (2007)
Proceedings of the Materials Forum 2007: Corrosion Education for the 21st Century (2007)
Managing Materials for a 21st Century Military (2007)
A Matter of Size: Triennial Review of the National Nanotechnology Initiative (2006)
Proceedings from the Workshop on Biomedical Materials at the Edge: Challenges in the Convergence of Technologies (2006)
Defending the U.S. Air Transportation System Against Chemical and Biological Threats (2006)
Globalization of Materials R&D: Time for a National Strategy (2005)
Going to Extremes: Meeting the Emerging Demand for Durable Polymer Matrix Composites (2005)
High-Performance Structural Fibers for Advanced Polymer Matrix Composites (2005)
Nanotechnology for the Intelligence Community (2005)

Preface

Over the last decade, government agencies, academic institutions, industry, and others have conducted many assessments of the environmental, health, and safety (EHS) aspects of nanotechnology. The results of those efforts have helped to direct research on the EHS aspects of engineered nanomaterials (ENMs). However, despite the progress in assessing research needs and despite the research that has been funded and conducted, developers, regulators, and consumers of nanotechnology-enabled products remain uncertain about the types and quantities of nanomaterials in commerce or in development, their possible applications, and their associated risks. To address those uncertainties, the Environmental Protection Agency asked the National Research Council to perform an independent study to develop and monitor the implementation of an integrated research strategy to address the EHS aspects of ENMs.

In this report, the Committee to Develop a Research Strategy for Environmental, Health, and Safety Aspects of Engineered Nanomaterials presents a conceptual framework for the proposed research strategy and identifies critical research gaps and tools needed to address them. The committee identifies high-priority research that needs to be undertaken in the short and long term and the resources needed. The report concludes with a discussion of mechanisms to ensure effective implementation of the committee's research strategy. In a subsequent report, the committee will evaluate research progress.

This report has been reviewed in draft form by persons chosen for their diverse perspectives and technical expertise in accordance with procedures approved by the National Research Council Report Review Committee. The purpose of the independent review is to provide candid and critical comments that will assist the institution in making its published report as sound as possible and to ensure that the report meets institutional standards of objectivity, evidence, and responsiveness to the study charge. The review comments and draft manuscript remain confidential to protect the integrity of the deliberative process. We thank the following for their review of this report: Nathan Baker, Pacific Northwest National Laboratory; Michael Ellenbecker, University of Massachusetts Lowell; Richard Flagan, California Institute of Technology; Robert Hurt, Brown University; Jacqueline Isaacs, Northeastern University; Jennifer Kuzma, University of Minnesota; Terry Medley, E.I. duPont de Nemours & Co.; James Murday, University of Southern California; Andre Nel, University of California, Los Angeles; Joanne Shatkin, CLF Ventures, Inc.; Robert Tanguay, Oregon State University; David Tirrell, California Institute of Technology; Jason Unrine, University of Kentucky; Paul Westerhoff, Arizona State University; and Yannis Yortsos, University of Southern California.

Although the reviewers listed above have provided many constructive comments and suggestions, they were not asked to endorse the conclusions or recommendations, nor did they see the final draft of the report before its release. The review of the report was overseen by the review coordinator, Richard B. Schlesinger, Pace University, and the review monitor, Julia M. Phillips, Sandia National Laboratories. Appointed by the National Research Council, they were responsible for making certain that an independent examination of the report was carried out in accordance with institutional procedures and that all review comments were carefully considered. Responsibility for the final content of the report rests entirely with the committee and the institution.

The committee gratefully acknowledges the following for making presentations to the committee: Lynn Bergeson, Bergeson & Campbell, P.C.; P. Lee Ferguson, Duke University; Richard Judson, Jeffrey Morris, and James Willis, U.S. Environmental Protection Agency; Subhas Malghan, U.S. Food and Drug Administration; Scott McNeil, Science Applications International Corporation; Giovanni Parmigiani, Harvard University; Paul Schulte, National Institute for Occupational Safety and Health; Justin Teeguarden, Pacific Northwest National Laboratory; Alan Tessier, National Science Foundation; Sally Tinkle, National Nanotechnology Coordination Office, and formerly with the National Institute of Environmental Health Sciences; and Jason Unrine, University of Kentucky.

The committee is also grateful for the assistance of National Research Council staff in preparing this report. Staff members who contributed to the effort are Eileen Abt, project director; James Reisa, director of the Board on Environmental Studies and Toxicology; Tina Masciangioli and Erik Svedberg, senior program officers; Keegan Sawyer, associate program officer; Keri Schaffer, research associate; Norman Grossblatt, senior editor; Mirsada Karalic-Loncarevic, manager, Technical Information Center; Radiah Rose, manager, editorial projects; Orin Luke, senior program assistant; and Tamara Dawson, program associate.

I would especially like to thank the members of the committee for their efforts throughout the development of this report.

Jonathan M. Samet, *Chair*
Committee to Develop a Research Strategy
for Environmental, Health, and Safety Aspects
of Engineered Nanomaterials

Contents

APPENDIXES

BOXES, FIGURES, AND TABLES

BOXES

FIGURES

TABLES

A Research Strategy for Environmental, Health, and Safety Aspects of **ENGINEERED NANOMATERIALS**

Summary

Nanotechnology relies on the ability to design, manipulate, and manufacture materials at the nanoscale.[1] The emerging field of nanotechnology has the potential to lead to substantial advances in many sectors—energy, medicine, electronics, and clean technologies, for example—while contributing to substantial economic growth. Engineered nanomaterials (ENMs) are already in industrial and consumer products, including drug-delivery systems, stain-resistant clothing, solar cells, cosmetics, and food additives. It is the nanoscale-specific properties of ENMs (for example, their electronic, optical, or chemical-reactive qualities) that are key to research and commercial applications.

The nanotechnology sector, which generated about $225 billion in product sales in 2009, is predicted to expand rapidly over the next decade with the development of new technologies that have new capabilities. The increasing production and use of ENMs may lead to greater exposures of workers, consumers, and the environment, and the unique scale-specific and novel properties of the materials raise questions about their potential effects on human health and the environment. In light of the rising use of ENMs, this report was motivated by the need for a research strategy to address critical gaps in knowledge related to the unique properties and environmental, health, and safety (EHS) risks of ENMs. Major challenges in developing such a strategy include

- Great diversity of nanomaterial types and variants.
- Lack of capabilities to monitor rapid changes in current, emerging, and potential ENM applications and to identify and address the potential consequences for EHS risks.
- Lack of standard test materials and adequate models for investigating EHS risks, leading to great uncertainty in describing and quantifying nanomaterial hazards and exposures.

To address these challenges, the Environmental Protection Agency (EPA) asked the National Research Council to perform an independent study to de-

[1]Nanoscale refers to materials on the order of one billionth of a meter.

velop and monitor the implementation of an integrated research strategy on EHS risks of ENMs. In response to EPA's request, the National Research Council convened the Committee to Develop a Research Strategy for Environmental, Health, and Safety Aspects of Engineered Nanomaterials, which prepared this report. The committee was charged to create a conceptual framework for EHS-related research, to develop a research plan with short-term and long-term research priorities, and to estimate resources needed to implement the research plan[2]. In a subsequent report, the committee will evaluate research progress.

In this report, the committee presents a strategic approach for developing the science and research infrastructure needed to address uncertainties regarding the potential EHS risks of ENMs. This approach begins with a discussion of the need for a research strategy. The committee next describes a new conceptual framework that structures its approach, focusing on emerging materials that may pose unanticipated risks, and on the properties of ENMs and their influence on hazards and exposure. The committee then identifies critical research gaps reflecting the elements of the framework, and the tools needed for addressing these gaps. Together with the conceptual framework and the identified gaps and tools, the committee develops the research portfolio, identifying where changes in course are needed, and where additional cross-cutting research would add value. Resources needed to implement the research priorities are identified. Last, the committee discusses the need for mechanisms to ensure implementation of the research strategy and evaluation of research progress that will be conducted in the subsequent report.

WHY IS ANOTHER STRATEGY FOR ENVIRONMENTAL, HEALTH, AND SAFTETY RESEARCH NEEDED?

As nanotechnology has burgeoned, questions about the possible risks posed by ENMs have been raised, fueled in part by the increased production, by a growing awareness that adequate methods are not available to detect and characterize the materials in the environment, and by recognition that the materials are in products or environments where exposures potentially can occur. In response to those concerns, there has been an increase over the last decade both in funding for research and in peer-reviewed publications addressing EHS effects of ENMs, in particular from the U.S. National Nanotechnology Initiative (NNI), "the government's central locus for the coordination of federal agency investments in nanoscale research and development" (NRC 2009).[3]

Over the last decade, many assessments of the potential EHS effects of nanotechnology have been conducted worldwide by government agencies (in-

---

[2]See Chapter 1 for the committee's statement of task.

[3]NRC (National Research Council). 2009. Review of the Federal Strategy for Nanotechnology-Related Environmental, Health, and Safety Research. Washington, DC: National Academies Press.

cluding the NNI), academic institutions, and industry (see Table 1-1). Those efforts have helped to translate and communicate information on the potential EHS effects of nanotechnology among researchers who are generating the scientific evidence, the businesses that use nanotechnology, the consumers who are using products with ENMs, and the various regulators who are overseeing ENMs. In the United States, the NNI has coordinated the efforts of regulatory and research agencies in identifying and addressing cross-agency research needs. The NNI guidance is complemented by agency-specific research strategies. In addition, the 2009 National Research Council review of the federal strategy highlighted the coordinating functions of the NNI and identified elements that are integral to a research strategy, including input from various stakeholders and mechanisms to ensure that the research strategy will be supported and funded. The 2009 report also identified limitations of the NNI approach. The NNI's 2011 Draft EHS Strategy addresses some of the limitations and further develops a framework for coordination among federal agencies and mechanisms to support the implementation of the strategy.[4]

Despite some progress in assessing research needs and in funding and conducting research, developers, regulators, and consumers of nanotechnology-enabled products remain uncertain about the variety and quantity of nanomaterials in commerce or in development, their possible applications, and any potential risks. There is insufficient connection and integration between generation of data and analyses on emergent risks and strategies for preventing and managing the risks.

Based on the committee's review of the current state of research and its relation to the needs of developers, regulators, and users of ENMs, three particular gaps are evident. First, little research progress has been made on some key topics, such as the effects of ingested ENMs on human health. Second, there is little research on the potential health and environmental effects of the more complex ENMs that are expected to enter commerce over the next decade. Third, system-integrative approaches are needed that can address all forms of ENMs based on their properties and an understanding of the underlying biologic interactions that determine exposure and risk. In spite of the need to provide more certain information on potential EHS risks, the gaps in understanding identified in many scientific workshops over the last decade have not been aggressively addressed with needed research. Common themes identified in workshops include the need for standardized materials, standardized methods to evaluate exposures, both in the workplace and in the environment, and harmonized methods for in vitro to in vivo validation in hazard assessments. In addition, rapidly evolving research approaches reflect an increasing emphasis on high-throughput screening and predictive modeling, both essential for managing the complexity of ENMs.

---

[4]The final version of the strategy was released in October 2011. However, because this report had already gone through peer review, the final version of the NNI EHS strategy was not reviewed or commented on by the committee.

Thus, there is a need for a research strategy that is independent of any one stakeholder group, that has human and environmental health as the primary focus, that builds on past efforts and is flexible in anticipating and adjusting to emerging challenges, and that provides decision-makers with timely, relevant, and accessible information.

THE COMMITTEE'S CONCEPTUAL FRAMEWORK

The diverse properties of nanomaterials make them challenging from the perspective of risk assessment. The variety of ENM types and the variation within types make it difficult to define their composition, structure, and properties without extensive characterization. The countless assemblages of atoms and structures and the plethora of inorganic and organic macromolecular coatings affect ENM surface chemistry and thus their behavior in the environment and in organisms. Depending on the environment where a nanomaterial is present (for example, lung fluid, surface water, or air), its surface properties may change, affecting its behavior, so that making predictions about such behavior and potential effects is challenging. Because of the variety of ENMs with differing properties, it is difficult to identify materials or classes of materials that may behave similarly with respect to fate, transport, toxicity, and risk.

In developing the new conceptual framework for considering ENM-related risks and for shaping the direction of the research portfolio (see Figure S-1), the committee considered properties that might be identified in a new nanomaterial that could pose a new, enhanced, or ill-defined risk. There is a need for an approach that promotes scientifically sound investigations of emerging risks and provides timely results relevant to the rapidly developing nanomaterial industry without relying on case-by-case evaluations of the nanomaterials.

The committee's conceptual framework is characterized by three key features:

- A value-chain[5] and life-cycle perspective that considers potential effects originating in the production and use of nanomaterials, nanomaterial-containing products, and the wastes generated.
- A focus on determining how nanomaterial properties (for example, size, surface characteristics, solubility, and crystallinity) affect key processes

[5]A value chain is a chain of activities that extends from the generation of nanomaterials to the production of primary and secondary products that are based on them. Activities along the value chain imply inputs of energy and materials at each stage and the creation of waste streams. For example, such ENMs as quantum dots (QDs) and single-walled carbon nanotubes (SWCNTs) might be combined as QD-SWCNT composites in primary products, such as thin films. Thin films might then be incorporated into solar cells (secondary products), which are then used in housing materials (tertiary products). All the products that form the value chain have their own life cycles associated with their manufacture, transport, processing, use, and end of life.

(for example, agglomeration, aggregation, dissolution, and deposition) that are relevant to predicting both hazard and exposure.

- The application of three principles that help guide our understanding of ENMs and research gaps when addressing ENM risks. The three principles address the emergent nature of ENM risks, the plausibility or likelihood of significant risks, and the potential severity of an effect.[6]

Figure S-1, which is not intended to show a linear process, depicts sources of nanomaterials originating throughout their value chains and life cycles and considers the environmental or physiologic context of these materials and the processes that they affect. The circle, identified as "critical elements of nanomaterial interactions," represents the physical, chemical, and biologic properties or processes that are considered most critical for assessing exposure, hazard, and hence risk. Those elements exist on many levels of biologic organization, including molecules, cells, tissues, organisms, populations, and ecosystems. The committee asks, What are the most important elements to examine to determine whether a nanomaterial is harmful? It has placed these elements at the center of the proposed research framework. The lower half of the figure depicts tools that can support a research agenda on the critical elements of nanomaterial interactions. The tools include materials (standardized ENMs that represent a variety of characteristics of interest), methods (standardized approaches for characterizing, measuring, and testing ENMs), models (for example, for assessing availability, concentration, exposure, and dose), and informatics[7] (methods and systems for systematically capturing, archiving, and sharing research results). The vertical arrows between the tools and the circle containing the critical elements represent the interplay between what is learned about the processes that influence exposure and hazard and the continuing evolution of the tools for carrying out research.

The committee's framework assumes that EHS research priorities can be determined on the basis of judgments regarding the relationships between nanomaterial *properties* and the *processes* that govern their interactions with organisms and ecosystems. These interactions will ultimately shed light on the emergent and plausible risks posed by the materials. Addressing gaps in our knowledge of these processes requires the recognition that many of the key research questions are systems issues—that is, they can be addressed only by considering the interactions of the various components along the life cycle of nano-

---

[6]The principles help in identifying nanomaterials that require closer scrutiny regarding risk, irrespective of whether they are established, emerging, or experimental ENMs. The principles also help to avoid a reliance on rigid definitions of ENMs.

[7]Informatics is defined here as the infrastructure and information science and technology needed to integrate data, information, and knowledge. An overall purpose of informatics in the context of EHS aspects of nanotechnology is to organize data so that they can be mined to determine how nanomaterial properties affect their exposure and hazard potential and overall risks to the environment and human health.

materials. For example, this framework considers the evaluation of hazard and the evaluation of exposure as occurring in concert, rather than sequentially.

The conceptual framework supports a strategic approach to nanotechnology-EHS research. Critical gaps in knowledge (Chapter 3) and the need for improved tools—including materials, methods, models, and informatics—to explore them (Chapter 4) figure prominently in identifying priorities for research (Chapter 5).

CRITICAL RESEARCH GAPS

Despite the substantial research already done on potential EHS risks posed by ENMs, critical gaps remain. The committee, using its collective judgment and informed by the research literature, identified the most pressing research gaps that need to be addressed to understand the potential environmental and human health effects of nanotechnology. The gaps, identified below, are discussed in the context of the source-to-outcome paradigm reflected in the committee's framework (Figure S-1); tracking the lifecycle of an ENM as it is incorporated into a product during manufacturing, transported and transformed by processes that may facilitate exposure to humans and organisms, made biologically available to organisms or ecosystems, and finally, assessing its potential effects on organisms and ecosystems.

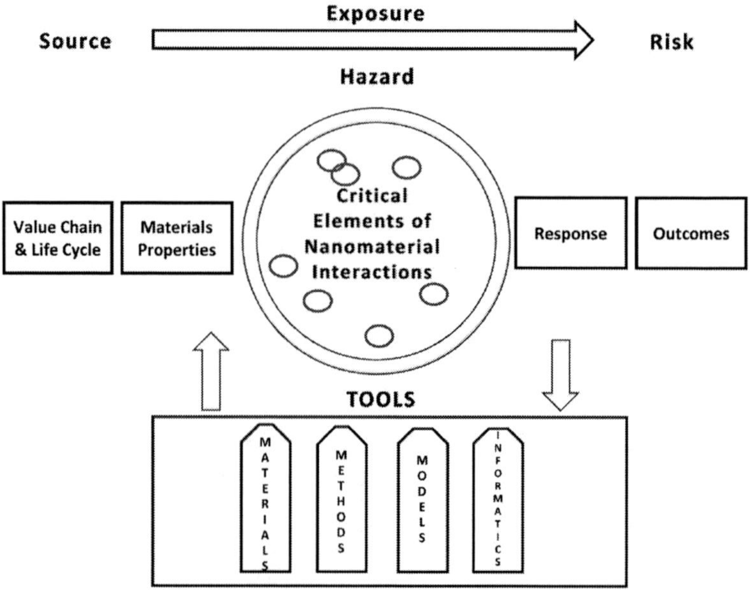

FIGURE S-1 Conceptual framework for informing the committee's research strategy.

The types of ENMs in products, the sources of exposure, and the expected magnitudes of the exposures typically are not known. Therefore, there is considerable uncertainty about potential exposures of populations—workers, consumers, and ecosystems. Because the nanotechnology market is projected to change rapidly, today's exposure scenarios may not resemble those of the future.

After identification of sources, exposures need to be assessed. Exposure assessment should include evaluation of modifications of ENMs across their lifecycle, as materials may undergo both subtle and extreme changes as they move through biologic and environmental systems that affect their size, surface chemistry, and reactivity.

Human exposures potentially occur through inhalation, oral, and dermal routes. Research gaps in understanding both general and occupational exposures persist.

- More is known about inhalation exposures because of past research on particles than about other routes of exposure. It is not clear, however, under what conditions airborne exposure to ENMs occur and what the exposure levels are likely to be, although application-specific processes could result in inhalation exposure.
- Little is known about dermal and ingestion exposures relevant to expected exposures of consumers to personal-care products and through food.

Little is known about the transport and distribution of ENMs in the human body and in the environment. When ENMs enter the human body, their surfaces may be modified by native biomolecules. Similarly, ENMs in the environment undergo transformations—for example, dissolution, aggregation, disaggregation, and chemical transformation.

- Research is needed to understand these biomolecular modifications in the human body.
- There is also a need to understand environmental transformation processes and their variation with ENM structure as these environmental modifications of ENMs can affect transport, fate, exposure, and toxicity.

After release throughout the life cycle, ENMs may enter the environment and reach organisms. The connection between the amount of an ENM at the interface with an organism and its relevant bioavailability is largely unknown. There are considerable uncertainties in understanding dose, biodistribution, and bioaccumulation of ENMs in humans and organisms.

- Doses used in biokinetic animal studies and for extrapolating from in vivo to in vitro studies need to be informed by relevant data on human exposures, whether in the workplace, in a laboratory, or in consumer use.

In the environment ENMs will persist or accumulate mainly in the solid and aqueous phases. Such environmental media may act as diluting agents, but only if the ENMs do not distribute and concentrate in particular compartments (for example, sediment or organisms).

- There is a need to understand the potential for ENMs to distribute into particular environments. This requires an ability to measure and characterize ENMs in different environmental media. Relative to human health exposure assessment, monitoring for environmental exposure to ENMs is in its infancy.

The responses of humans, other organisms, and ecosystems to ENMs are central to an understanding of risks. Most toxicity studies test a single material; however, there is incomplete information on effects of the array of ENMs used in products. Toxicologic studies usually focus on effects of acute exposures, and there is a lack of information on effects of chronic exposures.

Additional research is needed to understand potential human health risks from ENMs.

- Most ENM hazard assessments have relied on in vitro testing with doses that are often orders of magnitude higher than realistic exposures. It is important to understand what biologic effects occur at realistic ENM doses and dose rates and how ENM properties (for example surface coating) and exposure methods (for example, inhalation vs instillation) influence the magnitude of these effects.
- There is a need to develop data that addresses correlation between in vitro and in vivo responses. These data are vital for developing high-throughput screening strategies for ENMs.

There are considerable gaps in our understanding of the potential risks of ENMs to ecologic receptors. There are a large number of exposure routes and receptors, and the relationship among organism effects, population effects, and ecosystem responses are complex.

- Research is needed to guide identification of appropriate ecologic receptors, to develop appropriate ENM assays, and to conduct model ecosystem studies that address potential effects on a larger scale, such as a population, a community, or an ecosystem.
- Although numerous standard screening-level toxicity tests for specific aquatic and terrestrial organisms have been proposed for evaluating effects of ENMs, data are needed to assess whether standard tests can predict ecosystem effects of ENMs.

TOOLS AND APPROACHES NEEDED TO ADDRESS RESEARCH GAPS

To address these research gaps, the committee identified tools needed for characterizing how the properties of ENMs influence their biologic and environmental interactions (Figure S-1). Primary research needs related to tools include access to well-characterized nanomaterials; methods for characterizing, measuring, and testing materials in environmental and biologic samples and for assessing exposure and toxicity; exposure and effects modeling; and informatics.

Identifying ENM properties relevant to biologic and environmental interactions will require well-characterized libraries of materials for hypothesis-testing, as well as reference or standard test materials that may be used as benchmarks for making comparisons among studies, for validating protocols or measurements, or for testing specific hypotheses related to material properties and specific outcomes. Research or commercial materials are needed to study their biologic or ecologic effects, as these materials have the greatest potential to be released into the environment.

- The lack of widespread access to such materials and the lack of agreement as to which materials to consider as standards impede progress toward correlating ENM properties with their effects, make comparisons among studies difficult, and limit the utility of data gathered through informatics.
- The types of materials needed and used will depend on the purpose of the research. Each type of material needs to be characterized sufficiently for test results to be reproducible and for relationships between observed effects and material structure and composition to be defined.

Chemical and physical information on ENMs in environmental and biologic matrices is needed. Many existing analytic techniques in material sciences and other disciplines are applicable to ENMs, but their use in measuring and characterizing low concentrations and heterogeneous matrices will require additional development or, in some cases, development of new approaches.

- Tools and processes are needed for detecting, tracking, and characterizing ENMs in situ or in vivo at low concentrations; methods also are needed for assessing ENM transformations.

Protocols and techniques are needed for assessing the toxicity of ENMs.

- Existing toxicity-testing protocols for chemicals will need to be adapted or new methods will need to be developed and validated to include relevant cell types and organisms, appropriate dosimetry, and toxicity end points.
- Mechanistic data are needed for understanding the relevance of short-term high-dose effects to longer-term risks. Therefore, protocols should be developed and validated for extrapolating from short-term effects to long-term low-dose risks.
- In vitro and ultimately in silico toxicity-testing protocols need to be developed and applied to yield toxicity information that correlates with in vivo responses. This will require standardized and validated in vitro methods (for example, standardized cell types and exposure protocols) that represent specific

exposure routes and validation of results from in vitro studies against responses in relevant in vivo studies.

- Additional protocols are needed for predicting population or ecosystem effects of chronic ENM exposures of specific organisms and for assessing the indirect effects of ENMs, for example, their effects on carbon and nitrogen cycling.

To understand exposures to ENMs, standard testing protocols are needed to assess the properties that influence the transport, transformation, persistence, accumulation, and bioavailability of ENMs.

- These protocols need to be assessed and validated among various ENM types and classes and under various environmental conditions (for example, freshwater, seawater, and groundwater environments).
- Standard protocols and analytic methods that measure particle number, surface area, and mass concentration also are needed to assess airborne exposure to ENMs in epidemiologic and occupational studies.

The variety of ENM types and properties will require the development and use of models to predict exposure to ENMs and the effects of exposure; that is, models of exposure, bioavailability, mechanistic toxicity, biodistribution, and dosimetry.

Because of the paucity of data on the behavior of ENMs in organisms and in the environment and on the quantities of ENMs in environmental media, the development of more useful exposure models requires information regarding ENM sources, transport, transformations, fate, and bioavailability.

- Developing models for predicting releases of ENMs into the environment throughout their life cycle and value chain will require information on the types of materials being produced and used, types of applications, and intended uses.
- To understand the transport of the ENMs into the environment, existing exposure models need to be modified to include processes most relevant to ENM distribution in the environment and human exposure.
- Appropriate metrics for incorporating transformations and persistence into exposure models (for example, half-life time, size, or change in number concentration) need to be determined.

If in vitro assays are to be used as a predictive tool, mechanisms of toxicity of ENMs need to be a major research focus to assess whether mechanisms operative in vitro also apply in vivo.

- Models of effects should consider at least each of the four generally recognized mechanisms of toxicity: inflammation and oxidative stress, immunologic mechanisms, protein aggregation and misfolding, and DNA-damage mechanisms. However, there may be other mechanisms that have yet to be identified.

- For modeling ecologic effects, more data need to be collected on sublethal end points of toxicity, including effects on organism growth, behavior, reproduction, development, and metabolism.
- Data at the cellular or organism scale cannot predict effects at the community and ecosystem levels. Data should be collected on these effects (for example, on community structure and nutrient cycling) to determine potential model assays that can be used to improve prediction of chronic effects in a broad array of representative organisms and changes in ecosystem function.

With regard to dosimetric models for using exposure concentrations to predict dose, models are needed to determine biodistribution—including uptake, translocation, and elimination pathways and mechanisms, and to predict bioavailability of ENMs.

Informatics are needed to collect, analyze, and share the highly diverse set of data types and formats being generated to predict the potential exposure and effects of ENMs on the basis of their properties. Given the rapidity at which nanomaterials and their products are being introduced into commerce, an informatics infrastructure is needed to address the identified data gaps and support more efficient approaches for methods and model development and for data-sharing among the broad disciplines involved in nanotechnology research, development, translation, and regulation. However, optimal use of informatics requires collaboration among academe, industry, regulatory bodies, and others. The benefits of collaboration are numerous and include the sharing of data and models, the use of Web-based tools for rapid dissemination and communication between disciplines, and ultimately acceleration of research. However, there are scientific and technical barriers to the use of informatics, as well as organizational and cultural challenges that need to be overcome.

RESEARCH PRIORITIES AND RESOURCE NEEDS

Having considered the research gaps and the needed tools, the committee identified four broad, cross-cutting high-priority categories that form the backbone of its recommended research strategy. Because of the diversity of nanomaterials and the breadth of their potential applications, the committee considered that a prescriptive approach to addressing the EHS aspects of nanomaterials would be short-sighted and would probably fail to anticipate the rapid evolution of this field and its potential effects. Rather, in selecting the four broad categories, the committee envisions a risk-based system that is iteratively informed and shaped by research outcomes and that supports approaches to environmental and human-health protection even as our knowledge of ENMs is expanding and the research strategy is evolving.

The committee considers the four categories to be of equal priority and interrelated. This report describes aspects of these categories that need to be addressed in the short term (within 5 years)—on the grounds that these activities can be readily organized, resourced, and accomplished with available knowledge and tools—and others that will evolve over longer terms, which indicate the iterative nature of the research agenda.

The research categories are the following:

- *Identification, characterization, and quantification of the origins of nanomaterial releases.* Research in this category would develop inventories on ENMs being produced and used, identify and characterize the ENMs being released and the populations and environments being exposed, and assess exposures to measure the quantity and characteristics of materials being released and to model releases throughout their life cycle. Industry involvement will be needed for understanding trends in nanomaterial markets.
- *Processes that affect both potential hazards and exposure.* Research topics in this necessarily broad category would include the role of nanoparticle-macromolecular interactions in regulating and modifying nanoparticle behavior on scales ranging from genes to ecosystems; the effects of particle-surface modification on aggregation and nanoparticle bioavailability, reactivity, and toxicity potential; processes that affect nanomaterial transport across biologic or synthetic membranes; and the development of relationships between the structure of nanomaterials and their transport, fate, and effects. As an element of this research category, instrumentation and standard methods will need to be developed to relate ENM properties to their hazard and exposure potential and to determine the types and extent of ENM transformations in environmental and biologic systems.
- *Nanomaterial interactions in complex systems ranging from subcellular systems to ecosystems.* Research is unified by the need to understand how ENMs interact with complex systems, whether subcellular components, single cells, organisms, or ecosystems. Each level of these systems is complex with many embedded, interrelated processes that may interact in different ways (for example, synergistically or antagonistically) in response to ENMs. Examples of research in this category include efforts to understand the relationship between in vitro and in vivo responses; prediction of system-level effects, such as ecosystem functions (for example, nutrient cycling), in response to ENMs; and assessment of the effects of ENMs on the endocrine and developmental systems of organisms.
- *Adaptive research and knowledge infrastructure for accelerating research progress and providing rapid feedback to advance research.* This category of activities will help to integrate the research agenda and provide support for work in the other categories. Activities would include making characterized nanomaterials widely available, refining analytic methods continuously to define the structures of the materials throughout their lifespan, defining methods and

protocols to assess effects, and increasing the availability and quality of the data and models. Informatics would be fostered by the joining of existing databases, the encouraged and sustained curation and annotation of data, and the assignment of credit to those who share datasets and models.

The committee surveyed the status of existing resources needed to implement a strategic research plan within the context of these research-priority categories, and concluded that there is a gap between the research and associated activities funded and the level of activity that would foster greater progress toward providing information and tools to support the committee's research strategy. In considering how to address this gap, the committee took a pragmatic approach that was informed by its expert judgment based on the research priorities identified and knowledge of the cost of research activities, balanced with understanding of the current funding constraints. On the basis of this approach, the committee concludes that its strategy affords an opportunity for realignment of the substantial federal resources being dedicated to nanotechnology-related EHS research—$123.5 million in the president's FY 2012 budget request (5.8% of the total nanotechnology R&D budget). Such realignment will require federal-agency cooperation and resource leveraging.

However, infusion of modest additional resources could have a substantial effect on infrastructure that is critical for supporting an effective R&D program to advance the strategy. These additional resources will need to be garnered through a coordinated effort on the part of those involved with ENMs to leverage additional resources from public, private, and international initiatives to support critical cross-cutting research. Cross-cutting activities are encompassed in the high-priority research categories and will need to be supported by greater coordinated investment in five areas: nanotechnology-related EHS informatics; investment in translating advanced nanomaterial measurement and characterization approaches to EHS-relevant applications; investment in developing and providing benchmark nanomaterials; investment in identifying and characterizing nanomaterial sources throughout the value chain and life cycle of products; and investment in developing and maintaining research networks that provide human infrastructure for collaborative research, information-sharing, and translation of knowledge to effective use. Without budgetary increases in each of these areas, the committee anticipates that the federal government's ability to derive strategic value from investments in nanotechnology-related EHS research will remain insufficient.

Specifically, to ensure the development and implementation of the strategy,

- It is assumed that core EHS R&D funding by federal agencies should remain at about $120 million[8] per year over the next 5 years. Any reduction in

---

[8]This figure is an estimate from the president's FY2012 budget request of $123.5 million.

this total would be a setback to EHS research and would slow progress in addressing the committee's priorities.

- Over time, funded research should be aligned with the strategic priorities identified by the committee and in the NNI strategy.
- Additional multiagency funding should be made available for five cross-cutting endeavors that are critical for providing needed infrastructure and materials to support a strategic R&D program and for ensuring that research findings can be readily translated into practical action by stakeholders. The five are informatics ($5 million per year), instrumentation ($10 million per year), materials ($3-5 million per year), sources ($2 million per year), and networking ($2 million per year).[9]
- Funding in each of those five endeavors is critically needed in the short term and should be sustained for 5 years.

IMPLEMENTATION AND EVALUATION

To advance the research strategy, mechanisms will be needed to ensure its effective implementation, to evaluate research progress, and to refine the strategy as the base of evidence evolves—elements that the committee considered integral to its charge. Implementation will also require the integration of the various participants, both domestically and internationally, involved in nanotechnology-related EHS, including the NNI and the federal agencies; the private sector, such as nanomaterial developers and users; and the broader scientific and stakeholder communities, such as academic researchers.

Successful implementation will require mechanisms that improve coordination and modify institutional arrangements. Such modifications have been articulated by stakeholder groups involved in the nanotechnology-related EHS research enterprise. The committee concludes that attention to these implementation mechanisms are as integral to the success of the research strategy as the research priorities themselves, a key finding of the 2009 NRC review of the federal strategy. Active engagement of stakeholders is needed at all stages of strategy development, implementation, and revision to ensure that the research strategy is responsive to those who have a stake in its outcomes. Development of public-private partnerships can help to leverage resources to advance the research needed and to foster independent governance and operational transparency in the process. The committee considers that the current structure of the NNI, which serves primarily coordinating and information-sharing roles, hinders its accountability for effective implementation of the research strategy. Because the NNI has only coordinating functions, it has no "top-down" budgetary or management authority to direct nanotechnology-related EHS research. The committee finds that effective implementation of its strategy will require an entity that has sufficient management and budgetary authority to direct develop-

[9]The specified amounts are the minimums that should be available for each endeavor.

ment and implementation of a federal EHS strategy across NNI agencies and to ensure integration of federally supported EHS research with research undertaken by the private sector, the academic community, and international organizations.

There is a concern that the dual and potentially conflicting roles of the NNI—developing and promoting nanotechnology and its applications while identifying and mitigating risks that arise from such applications—impede implementation and evaluation of the EHS risk research. That duality is reflected in the diverse missions of the agencies and departments that make up the NNI. Numerous stakeholders have called for a separation of the two roles in the NNI, and such separation has historical precedent. To implement the research strategy effectively, a clear separation of management and budgetary authority and accountability is needed between the functions of developing and promoting applications of nanotechnology and of understanding and assessing potential health and environmental implications. Such a separation is needed to ensure that progress in implementing an effective nanotechnology-related EHS research strategy is not hampered. The separation of management of applications-targeted from management of implications-targeted research needs to be achieved through means that do not impede the free flow of ideas and results between the two lines of research.

In its second report, the committee will assess progress in understanding the EHS aspects of nanotechnology and the extent to which high-priority research has been initiated or implemented. The timeframe for the completion of the second report is too short to have substantial new research programs, let alone research outcomes, in place. But the committee considers that it is sufficient to see progress in initiating research in each of the four high-priority categories and progress in developing the infrastructure, accountability, and coordination mechanisms needed for implementation of the strategy. Progress in addressing those foundational elements will go a long way toward ensuring effective support and management of the research needed to provide information for identifying, assessing, and managing the potential EHS consequences of ENMs.

CONCLUDING REMARKS

Despite the promise of nanotechnology, without strategic research into emergent risks associated with it—and a clear understanding of how to manage and avoid potential risks—the future of safe and sustainable nanotechnology-based materials, products, and processes is uncertain. In today's fast-paced and interconnected world, a worthwhile economic and social return on government and industry investment in nanotechnology is unlikely to be fully realized without research on risk, including research on translation of knowledge into evidence-informed and socially responsible decision-making.

1

Background

OVERVIEW

On January 21, 2000, President Clinton announced a new U.S. initiative to explore and exploit the science and technology of matter on the nanometer scale (often referred to as the nanoscale). In an address at the California Institute of Technology on science and technology, President Clinton asked his audience to imagine "materials with 10 times the strength of steel and only a fraction of the weight; shrinking all the information at the Library of Congress into a device the size of a sugar cube; detecting cancerous tumors that are only a few cells in size." The speech laid the foundation for the National Nanotechnology Initiative (NNI) (The White House 2000). The NNI—"the government's central locus for the coordination of federal agency investments in nanoscale research and development" (NRC 2009, p. 3)—has set the pace for national and international research and development in nanoscale science and engineering and has led the world in the development and use of knowledge at the nanoscale with the potential to improve quality of life, stimulate economic growth, and address many of society's most pressing challenges.

With our understanding of the role that nanoscale science and engineering can play in the development of innovative materials, processes, and products has also come the knowledge that nanotechnology may lead to new mechanisms by which people and the environment may be harmed. In today's complex, interconnected, and resource-constrained world, it is important that products resulting from novel and emerging technologies that have uncertain risks, such as nanotechnology, be developed responsibly; that all stakeholders have an active role in socially responsible development; and that potential risks are identified and avoided as early as possible during research, innovation, and commercialization. This report maps out a research strategy that is intended to promote the responsible development of nanotechnology-enabled materials, processes, and products; and it offers an approach for helping to ensure that researchers, manufacturers, regulators, and others have the necessary information on potential risks and how to prevent, avoid, or mitigate them.

OPPORTUNITIES AND CHALLENGES

Increasing our understanding of how matter on the nanoscale behaves and interacts with humans and ecologic systems is socially and economically important. In a world where the needs of a growing population threaten to outstrip increasingly limited resources and many global challenges remain unresolved—from disease to hunger to renewable energy—nanotechnology, along with other fields of technologic innovation, can contribute to a sustainable future (Maynard 2010). Nanoscale science and technology are leading to new ideas and tools that can enhance existing technologies and create new ones, help to support new jobs, revitalize economies, and contribute to solutions to some of society's most pressing problems. But investing in research and development is just one step toward ensuring socially responsible, relevant, and successful technologic innovation. Realizing the economic and societal benefits of nanotechnology also requires educating the workforce, lowering barriers to technology transfer, and engaging with diverse stakeholders. And success with nanotechnology will also depend on developing and implementing new approaches to risk prevention and risk management that avoid past mistakes, that address issues in the innovation process, and that develop materials responsibly without impeding innovation unduly.

As nanotechnology research and development have led to new materials—nanomaterials—questions about the safety of these materials have prompted concerns that they are likely to be attended by new risks. Specifically, concerns have been raised that materials behaving in unconventional ways might lead to unanticipated risks to human health and the environment. Those concerns were underpinned and to an extent driven by research in the 1990s that showed that inhaled fine particles have the potential to cause more serious health effects than those estimated in studies of larger particles (for example, Oberdörster et al. 2007). The research signaled the beginning of a paradigm shift away from an understanding that risk stems from chemical composition alone to a recognition that physical form and chemical properties are both important for understanding, predicting, and preventing harm.

The concerns were exacerbated by the increase in production of materials that behaved in unique ways *because of their physical form on the nanoscale* and by growing awareness that methods for detecting, characterizing, monitoring, or controlling these materials in the environment were not available and that the materials were in products or in environments in which human exposures could occur (for example, see Maynard et al. 2006). Consequently, there is uncertainty about the potential human health and environmental effects of products emerging from nanotechnology and recognition that the safe and successful development of nanotechnology depends on early, strategic action to address potential risks.

In response to the concerns, there has been an increase in funding for research and in peer-reviewed publications addressing the environmental, health, and safety (EHS) effects of engineered nanomaterials (ENMs) (PCAST 2010).

In FY 2005, the combined investment by U.S. federal agencies in research on and development of EHS implications of nanotechnology was $34.8 million (NSET 2006). In FY 2012, the President's budget request proposes $123.5 million—more than a threefold increase (NSET 2010). Worldwide publications addressing the EHS effects of ENMs have increased similarly, with 791 papers published in 2009 compared with 179 publications in 2005 (PCAST 2010).

In 2006, the NNI published the first U.S. interagency assessment of EHS research needs associated with ENMs, identifying 75 research needs in five broad categories (NEHI 2006). The needs were assigned priorities by the Nanotechnology Environmental Health Implications Working Group (NEHI 2007) and were incorporated into an interagency research strategy by NEHI in 2008 (NEHI 2008). Recently, NEHI published a draft update (NEHI 2010) of the interagency research strategy that responds to input from the President's Council of Advisors on Science and Technology, a National Research Council review of the 2008 NEHI report (NRC 2009), and various stakeholder groups, including members of the public.[1] Those documents and many similar and complementary assessments by government agencies, academic institutions, industry, and other stakeholders (see Table 1-1) have helped to direct where EHS research should be focused if ENMs are to be developed and used safely. Yet despite progress in the development of research needs and in the amount of research that is funded and conducted, developers, regulators, and consumers of nanotechnology-enabled products remain uncertain about the types and quantity of nanomaterials in commerce or in development, their possible applications, and the potential risks associated with them.

It is the disconnect between risk research and its relevance to and use in informed decision-making that prompts the question, How can research best be guided and conducted to ensure that the products of nanotechnology are developed as safely, responsibly, and beneficially as possible? That question is central to the charge to this committee.

COMMERCIALIZATION OF ENGINEERED NANOMATERIALS

The development and use of new materials cannot be separated from questions of potential risk. Understanding and addressing the EHS implications of ENMs is intricately entwined with their development.

Over the last few years, industries—ranging from electronics to energy, materials to medicine, and chemicals to clean technologies—have been using nanotechnology to develop breakthrough innovations for products. To respond to the many opportunities, a global network of large corporations, academic

[1]A final version of the strategy was published in October 2011 (NEHI 2011). Because the committee's report had already gone to peer review, NEHI 2011 was not reviewed by this committee.

TABLE 1-1 Key Reports That Assess or Provide Information on Research Needs and Strategies for Addressing the Environmental, Health, and Safety Implications of Engineered Nanomaterials[a]

Year	Report	Source	Relevance
2004	Nanoscience and Nanotechnologies: Opportunities and Uncertainties	RS/RAE 2004	Identifies strategic-research gaps
2004	Technological Analysis: Industrial Application of Nanomaterials - Chances and Risks	Luther 2004	Identifies strategic-research gaps
2004	Nanotechnology: Small Matter, Many Unknowns	SwissRe 2004	Identifies strategic-research gaps
2005	Characterizing the Potential Risks Posed by Engineered Nanoparticles: A First UK Government Research Report	DEFRA 2005	Identifies strategic-research gaps
2005	Communication from the Commission to the Council, the European Parliament and the Economic and Social Committee. Nanoscience and Nanotechnologies: An Action Plan for Europe 2005 - 2009.	EC 2005	Identifies strategic-research gaps
2005	A Proposal to Increase Federal Funding of Nanotechnology Risk Research to at least $100 Million Annually	Denison 2005	Identifies strategic-research gaps
2005	Joint NNI-ChI CBAN and SRC CWG5 Nanotechnology Research Needs Recommendations	Vision 2020/SRC 2005	Identifies strategic-research gaps
2005	Small Sizes that Matter: Opportunities and Risks of Nanotechnologies	Allianz/OECD 2005	Provides contextual information on strategic risk research
2006	Opinion on the Appropriateness of Existing Methodologies to Assess the Potential Risks Associated with Engineered and Adventitious Products of Nanotechnologies	SCENIHR 2006	Provides contextual information on strategic risk research
2006	Nanotechnology: A Research Strategy for Addressing Risk	Maynard 2006	Outlines a research strategy
2006	Safe handling of nanotechnology	Maynard et al. 2006	Identifies strategic-research gaps
2006	Characterizing the Environmental, Health and Safety Implications of Nanotechnology: Where Should the Federal Government Go From Here?	ICFI 2006	Provides contextual information on strategic risk research
2006	White paper on Nanotechnology Risk Governance	Renn and Roco 2006	Provides contextual information on strategic risk research
2006	Environmental, Health and Safety Research Needs for Engineered Nanoscale Materials	NEHI 2006	Identifies strategic-research gaps
2007	Opinion on the Appropriateness of the Risk Assessment Methodology in Accordance with the Technical Guidance Documents for New and Existing Substances for Assessing the Risks of Nanomaterials	SCENIHR 2007	Identifies strategic-research gaps

(Continued)

TABLE 1-1 Continued

Year	Report	Source	Relevance
2007	Opinion on Safety of Nanomaterials in Cosmetic Products	SCCP 2007	Provides contextual information on strategic risk research
2007	Nano Risk Framework	EDF/DuPont 2007	Provides contextual information on strategic risk research
2007	Nanotechnology White Paper	EPA 2007	Identifies strategic-research gaps
2007	Nanotechnology: A Report of the U.S. Food and Drug Administration Nanotechnology Task Force	FDA 2007	Provides contextual information on strategic risk research
2007	Nanotechnology Recent Development, Risks and Opportunities	Lloyds 2007	Provides contextual information on strategic risk research
2007	Prioritization of Environmental, Safety and Health Research Needs for Engineered Nanoscale Materials: An Interim Document for Public Comment	NEHI 2007	Identifies strategic-research gaps
2007	Characterizing the Potential Risks Posed by Engineered Nanoparticles: A Second UK Government Research Report.	DEFRA 2007	Identifies strategic-research gaps
2007	Meeting Report: Hazard Assessment for Nanoparticles—Report from an Interdisciplinary Workshop	Balbus et al. 2007	Identifies strategic-research gaps
2008	Proceedings of the Workshop on Research Projects on the Safety of Nanomaterials: Reviewing the Knowledge Gaps	Höck 2008	Identifies strategic-research gaps
2008	Small is Different: A Science Perspective on the Regulatory Challenges of the Nanoscale	Council of Canadian Academies 2008	Identifies strategic-research gaps
2008	Engineered Nanoscale Materials and Derivative Products: Regulatory Challenges	Schierow 2008	Provides contextual information on strategic risk research
2008	Nanotechnology: Better Guidance is Needed to Ensure Accurate Reporting of Federal Research Focused on Environmental, Health and Safety Risks	GAO 2008	Provides contextual information on strategic risk research
2008	Responsible Production and Use of Nanomaterials	VCI 2008	Identifies strategic-research gaps
2008	Towards Predicting Nano-Biointeractions: An International Assessment of Nanotechnology Environment, Health, and Safety Research Needs	ICON 2008	Identifies strategic-research gaps
2008	Strategic Plan for NIOSH Nanotechnology Research and Guidance: Filling the Knowledge Gaps. Draft Report	NIOSH 2008	Outlines a research strategy

2008	Strategy for Nanotechnology-Related Environmental, Health and Safety Research.	NEHI 2008	Outlines a research strategy
2008	Novel Materials in the Environment: The Case of Nanotechnology	RCEP 2008	Identifies strategic-research gaps
2009	Risk Assessment of Products of Nanotechnologies	SCENIHR 2009	Identifies strategic-research gaps
2009	Scientific Opinion: The Potential Risks Arising from Nanoscience and Nanotechnologies on Food and Feed Safety.	EFSA 2009	Provides contextual information on strategic risk research
2009	Workplace Exposure to Nanoparticles	Kaluza et al. 2009	Provides contextual information on strategic risk research
2009	Nanomaterial Research Strategy	EPA 2009	Outlines a research strategy
2009	Securing the Promise of Nanotechnologies: Towards Transatlantic Regulatory Cooperation	Breggin et al. 2009	Provides contextual information on strategic risk research
2009	Review of the Federal Strategy for Nanotechnology-Related Environmental, Health, and Safety Research	NRC 2009	Identifies strategic-research gaps
2009	EMERGNANO: A Review of Completed and Near Completed Environment, Health and Safety Research on Nanomaterials and Nanotechnology	Aitken et al. 2009	Identifies strategic-research gaps
2009	FAO/WHO Expert Meeting on the Application of Nanotechnologies in the Food and Agriculture Sectors: Potential Food Safety Implications	FAO/WHO 2009	Provides contextual information on strategic risk research
2010	ENRHES Engineered Nanoparticles: Review of Health and Environmental Safety	Stone et al. 2010	Identifies strategic-research gaps
2010	Nanotechnology: Nanomaterials Are Widely Used in Commerce, but EPA Faces Challenges in Regulating Risk	GAO 2010	Provides contextual information on strategic risk research
2010	Nanotechnologies and Food	UKHL 2010	Identifies strategic-research gaps
2010	UK Nanotechnologies Strategy: Small Technologies, Great Opportunities	HM Government 2010	Outlines a research strategy
2010	Report to the President and Congress on the Third Assessment of the National Nanotechnology Initiative	PCAST 2010	Provides contextual information on strategic risk research
2010	Nanotechnology Research Directions for Societal Needs in 2020: Retrospective and Outlook, Chapter 4	Nel et al. 2010	Outlines a research strategy
2010	National Nanotechnology Initiative 2011 Environmental, Health, and Safety Strategy[b]	NEHI 2010	Outlines a research strategy

[a]Reports are classified as either providing contextual information on strategic-risk research, identifying strategic-research gaps, or outlining a research strategy. With few exceptions (included for historical significance), these reports represent the assessments, opinions, and recommendations of panels of experts.

[b]A final version of the strategy was published in October 2011 (NEHI 2011).

laboratories, government-funded research centers, technology incubators, and startup companies has emerged to facilitate collaboration in technologic development. Those developers play unique roles in the three-stage nanotechnology value chain: production of ENMs, raw materials that make up the first stage of the value chain; primary products (also termed intermediate or nanointermediate products) that either contain ENMs or have been constructed from other materials to possess nanoscale features and that comprise the second stage; and secondary products, finished goods that incorporate ENMs or intermediate products—the third stage.

In 2009, developers generated $1 billion from the sale of nanomaterials, which constitute the initial stage of the nanotechnology value chain (Lux Research in Wray 2010). In that year, the top 11 classes of nanomaterials (based on production volume) were ceramic nanoparticles, carbon nanotubes, nanoporous materials, graphene, metal nanoparticles, nanoscale encapsulation, fullerenes, dendrimers, nanostructured metals, nanowires, and quantum dots (Bradley 2010). Of these classes of nanomaterials, ceramic nanoparticles (50%), carbon nanotubes (20%), and nanoporous materials (20%) accounted for 90% of the total production volume (about 3,500 tons) (Bradley 2010). The market for products that rely on nanomaterials is expected to grow to $3 trillion by 2015 (Lux Research 2008a,b).[2] Although the relative percentages (based on production volume) are not expected to change drastically through 2015, the more exotic materials, such as nanowires and quantum dots, are likely to experience the biggest jump in production because they are starting from a much lower baseline than older classes of materials, such as ceramic nanoparticles (Bradley 2010).

The nanomaterials that make up the first stage of the value chain are used in the development of primary products or nanointermediates. In 2009, developers generated $29 billion for this stage of the value chain (Lux Research in Wray 2010). The top 10 nanointermediate classes developed were coatings, composites, catalysts, drug-delivery systems, energy storage, sensors, displays, memory, solar cells, and filters (Bradley 2010). For example, a variety of ceramic nanoparticles can be added to coating formulations to enhance function, including antiscratch, antifriction, and antimicrobial properties. Coatings, composites, and catalysts are the most prevalent of the nanointermediates.

Nanointermediates are expected to generate about $480 billion in 2015 (Lux Research 2009a; Sibley 2009). Although the coatings, composites, and catalysts will make up a large share of the nanointermediate market in 2015 and beyond, they will be joined increasingly by intermediates in the health-care and life-sciences sectors, especially drug-delivery devices, and by intermediates in the electronics sector, that is, in logic chips and in memory applications.

Both nanomaterials and nanointermediates feed into nanoenabled products, the last stage of the value chain. In 2009, producers generated $224 billion from the sale of nanoenabled products (Lux Research in Wray 2010). The top 10 product classes developed were automotive products, buildings and construc-

---

[2]This figure includes nanomaterials, intermediate products, and nanoenabled products.

tion, consumer electronics, personal care, marine, aerospace, sporting goods, food and agriculture, industrial equipment, and textiles (Bradley 2010). Of automotives, for example, ceramic nanoparticle-based coatings are added to engines to increase fuel economy, and carbon nanotubes are added to fuel lines to reduce the risk of fire. In the future, the overall value of nanoenabled products is projected to reach about $2 trillion by 2015 (Lux Research in Wray 2010; Hwang and Bradley 2010; Lux Research 2009b), with automotives, buildings and construction, and consumer electronics dominating this stage of the value chain.

PRESENT STATE OF STRATEGIC NANOTECHNOLOGY ENVIRONMENTAL, HEALTH, AND SAFETY RESEARCH

Despite extensive investment in nanotechnology and increasing commercialization over the last 10 years, uncertainties about the EHS effects of nanomaterials and nanomaterial-based products remain. There continues to be a lack of clarity about the safety, regulatory, and governance challenges that need to be addressed if materials, processes, and products based on nanotechnology are to be developed responsibly. Research is slowly being translated and communicated between people who are generating new knowledge and those who hope to use it—regulators, businesses, and consumers. But in spite of the EHS research that is being conducted, there remains a lack of an overarching, priority-set, coordinated research strategy that encompasses stakeholders in the public and private sectors and that identifies critical questions, maps out a path to answers, and ensures accountability for providing decision-makers with timely information.

Because nanotechnology is presenting new and unusual challenges, moving from reactive to proactive research represents a substantial paradigm shift in how risk-related research is conducted. The specific EHS issues raised by ENMs remain difficult to address. Nevertheless, the NNI has succeeded in coordinating regulatory and research agencies in identifying cross-agency research needs and in beginning to address the needs. The NNI documents (NEHI 2006, 2007, 2008, and 2010) that set forth the federal government's research strategy for addressing EHS implications of ENMs are a considerable achievement and help to identify further research that is needed. Those documents are complemented by agency-specific research strategies (described below). However, despite the progress that has been made toward developing an interagency research strategy, research on the potential EHS implications of ENMs still lacks context with respect to what is already known, what is occurring now, and what is likely to be important in the future. In the absence of a clear and implementable research strategy, research appears to be driven predominantly by assumptions of what is important and what is scientifically interesting rather than by a clear, rationale assessment of what is needed.

The current need for an overarching research strategy has been responded to in part by *National Nanotechnology Initiative 2011 Environmental, Health, and Safety Strategy* (NEHI 2010). The closing chapter begins to develop a framework that will support coordination among federal agencies; including establishment of a set of principles to encourage agencies to work together productively and mechanisms that support the NEHI and the National Nanotechnology Coordination Office (NNCO) in implementing the strategy.

The principles in NEHI (2010) are designed to help to set priorities among nanomaterials; to establish systems for conducting reproducible, valid, and translatable research; to ensure that the resulting data are of high quality; to couple research to different risk-assessment needs; and to support partnerships with stakeholders and engagement with the international community. The principles set the scene for ensuring that relevant and responsive research is conducted rather than dictating which agencies conduct what research. However, there remains the task of combining the emerging framework with a strategically responsive research strategy that facilitates the generation and application of timely and relevant data and that extends beyond the borders of federal agencies to include engagement with stakeholders and the international community.

HISTORY OF NANOTECHNOLOGY ENVIRONMENTAL, HEALTH, AND SAFETY RESEARCH ASSESSMENTS

In the early 2000s, a number of reports from diverse sources began to question the safety of nanotechnology-enabled products and emerging ENMs. They ranged from the ETC Group's call for a moratorium on nanotechnology research (ETC Group 2003) to Sun Microsystems founder Bill Joy's influential article in *Wired* magazine questioning the dangers of emerging technologies (Joy 2000) to the reinsurance giant Swiss Re's assessment of the uncertain impacts of nanotechnology (Swiss Re 2004). In 2004, those concerns were placed into context by the UK Royal Society and Royal Academy of Engineering (RS/RAE) (RS/RAE 2004). After an extensive consultation and assessment period led by a panel of experts, the RS/RAE published a highly influential report examining the opportunities and uncertainties of "nanoscience and nanotechnologies." Concluding that "the lack of evidence about the risks posed by manufactured nanoparticles and nanotubes is resulting in considerable uncertainty" (RS/RAE 2004, p. 85), the report made recommendations for avoiding potential risks and developing a better understanding of them. Although the RS/RAE report's recommendations were directed toward the UK government, they were used worldwide to ground discussions of the responsible development of nanotechnologies based on the state of knowledge about potential risks.

A number of risk-research assessments followed. In the UK, a response to the RS/RAE report by the UK government began to map out strategic research needs and a plan of action to address them (DEFRA 2007), and the European Union (EU) incorporated nanotechnology EHS research and development into its action plan (2005-2009) for developing nanoscience and nanotechnology. Those activities were complemented by the EU Scientific Committee on Emerging and Newly Identified Health Risks (SCENIHR) publication of a comprehensive review of "engineered and adventitious products of nanotechnologies," which is one of the most comprehensive reviews of the potential health risks associated with ENMs (SCENIHR 2006).

In December 2004, the NNI published its first strategic plan, which included a specific commitment to address the potential EHS implications of nanotechnology (NSET 2004). Under the goal of supporting the responsible development of nanotechnology, the NNI committed to "as necessary, expand support for research into environmental and health implications of nanotechnology" (NSET 2004, p. 12).

The strategic plan also stated (NSET 2004, p. 12) that

> NNI-funded research will (1) increase fundamental understanding of nanoscale material interactions at the molecular and cellular level through in vitro and in vivo experiments and models; (2) increase fundamental understanding of nanoscale material interactions with the environment; (3) increase understanding of the fate, transport, and transformation of nanoscale materials in the environment and their life cycles; and (4) identify and characterize potential exposure, determine possible human health impact, and develop appropriate methods of controlling exposure when working with nanoscale materials.

Federal activities were to be coordinated by NEHI, a group of representatives of research and oversight agencies that began meeting informally in 2003 and was formally established in 2004 (NSET 2004).

In 2005, in a parallel development, the NNI established two Consultative Boards for Advancing Nanotechnology (CBAN), consisting of representatives of the chemical and semiconductor industries, to help to define research needs. Those boards assembled subgroups that drew on public- and private-sector experts to address the potential EHS implications of nanotechnology and published joint research recommendations (Vision 2020/SRC 2005). Five key research needs were proposed by the joint groups: a testing strategy for assessing toxicity, best metrics for assessing particle toxicity, exposure-monitoring methods, risk-assessment methods, and communication and education concerning EHS and societal impacts (Ford 2005).

In keeping with those key needs, specific research topics were developed. For example, more detailed recommendations for near-term research to address the toxicity of nanomaterials were offered in an associated document and reflected discussions between government and industry representatives (NNI-ChI CBAN ESH Working Group 2007).

In 2006, Maynard (a member of the present committee) wrote a report that evaluated nine sources of information on EHS research needs, including

RS/RAE (2004) and CBAN (NNI-ChI CBAN ESH Working Group 2007), and outlined a possible federal research strategy (Maynard 2006). The report identified 11 subjects on which further research was needed: human-health hazards, health outcomes, environmental impact, exposure, material characterization, exposure control, risk reduction, standards, safety, informatics, and effective approaches to research. Priorities were set among specific research needs associated with those subjects for generating new knowledge and supporting oversight of emerging materials.

In November 2006, "Safe Handling of Nanotechnology" was published (Maynard et al. 2006). The coauthors were experts in government, industry, and the nongovernment sector. It established five overarching research challenges to developing nanotechnology-based products safely and proposed a timeline for the international research community to address them. The five challenges were to develop "instruments to assess exposure to ENMs in air and water, within the next 3-10 years"; "validated methods to evaluate the toxicity of ENMs, within the next 5–15 years"; "models for predicting the potential impact of ENMs on the environment and human health, within the next 10 years"; "robust systems for evaluating the health and environmental impact of ENMs over their entire life, within the next five years"; and "strategic programmes that enable relevant risk-focused research, within the next 12 months."

In that same year, the 2006 NEHI report was published. It categorized the 75 nanotechnology-related EHS research needs in five broad categories: instrumentation, metrology, and analytic methods; nanomaterials and human health; nanomaterials and the environment; health and environmental surveillance; and risk-management methods. The report was alluded to by Clayton Teague, director of NNCO, in testimony to the House Committee on Science and Technology on November 17, 2005, in which he noted that "a carefully designed research plan, along with shared Government and industry responsibility and collaboration should guide our efforts" (Teague 2005, p. 27). Following publication of NEHI (2006), a consultative document that presented 25 research priorities (five in each of the five research categories identified in the NEHI report) was produced (NEHI 2007).

In February 2008, the NNI strategy for nanotechnology-related EHS research was published (NEHI 2008). Building on the previous two reports (NEHI 2006, 2007), it expanded on the 25 research priorities identified in 2007 and indicated the changing emphasis that they should receive in the near, middle, and long terms in a research strategy. The document also developed a broad framework to guide strategic risk research and identified lead agencies for addressing research needs. The strategy was developed predominantly by government-agency representatives but drew on feedback from public consultations and responses to earlier documents.

The 2008 federal research strategy for nanotechnology-related EHS research (NEHI 2008) was reviewed by the National Research Council (NRC 2009), whose committee concluded (p. 10) that

> The NNI's 2008 *Strategy for Nanotechnology-Related Environmental, Health, and Safety Research* could be an effective tool for communicating the breadth of federally supported research associated with developing a more complete understanding of the environmental, health, and safety implications of nanotechnology. It is the result of considerable collaboration and coordination among 18 federal agencies and is likely to eliminate unnecessary duplication of their research efforts. However, the document does not describe a strategy for nano-risk research. It lacks input from a diverse stakeholder group, and it lacks essential elements, such as a vision and a clear set of objectives, a comprehensive assessment of the state of the science, a plan or road map that describes how research progress will be measured, and the estimated resources required to conduct such research.

Central to the stated concerns of that committee was a lack of an assessment of the state of the science, including previous efforts to identify and address strategic research needs, a research-gap analysis that overstated the scope and extent of relevant federally funded research, and a lack of stakeholder input into the process. The federal strategic plan has recently been substantially updated, on the basis of input from a series of workshops, stakeholder consultations, and reviews, including NRC (2009).

In parallel with the NNI efforts (and in part coordinated with them), individual federal agencies developed their own internal research strategies. The National Institute for Occupational Safety and Health (NIOSH) first published a draft EHS research strategy in 2005 (NIOSH 2005), and the most recent iteration was released in 2010 (NIOSH 2010). The current NIOSH strategy addresses four strategic goals, developed in consultation with stakeholders: "determine if nanoparticles and nanomaterials pose risks for work-related injuries and illnesses" (p. 18); "conduct research to prevent work-related injuries and illnesses by applying nanotechnology products" (p. 24); "promote healthy workplaces through interventions, recommendations, and capacity building" (p.24); and "enhance global workplace safety and health through national and international collaborations on nanotechnology research and guidance" (p. 27). Each goal is accompanied by specific subgoals and performance measures.

The Environmental Protection Agency (EPA) also developed a comprehensive nanotechnology EHS research strategy (EPA 2009). Building on a white paper first published in draft form in 2005 (EPA 2005), the strategy—made final in 2009—focuses on four research themes: "identifying sources, fate, transport, and exposure; understanding human health and ecological effects to inform risk assessments and test methods; developing risk assessment approaches; and preventing and mitigating risks" (EPA 2009, p. 1).

Those themes are addressed in the context of a decision-making framework to aid in identifying important research questions and an environmental-assessment approach based on Davis and Thomas's Comprehensive Environmental Assessment (Davis and Thomas 2006). For each theme, the strategy develops key scientific questions and outlines critical paths for the agency to follow to address them. Both the NIOSH and EPA strategies clarify their complementary relationships with NEHI (2008).

The agency-specific strategies have been complemented by a growing number of reviews and reports that identify nanotechnology EHS research needs, place nanotechnology EHS research within a broader context, and flesh out research strategies (see Table 1-1 for an illustrative list). In the UK, various organizations have addressed nanotechnology EHS issues in the wake of RS/RAE (2004), including the Royal Commission on Environmental Pollution (RCEP 2008) and the UK House of Lords (UKHL 2010). The UK government strategy for nanotechnology, published in 2010, places a high priority on understanding and addressing the potential health and environmental implications of ENMs (HM Government 2010).

In 2009, the Institute of Occupational Medicine published a comprehensive review of completed and nearly completed nanomaterial EHS research conducted worldwide (Aitken et al. 2009). Using a weight-of-evidence appraisal of over 260 research projects, the report assessed the current state of research and identified where information was lacking on the safety of ENMs. Seventy-one research gaps, covering all aspects of nanomaterial EHS effects, were identified.

In 2007-2009, the International Council on Nanotechnology hosted a series of three workshops to assess research needs related to EHS aspects of nanotechnology (ICON 2010a). The workshops brought together experts of various backgrounds, countries, and organizations to identify critical research needs in three categories: understanding principles that relate nanomaterial properties to defined risk factors, working toward predicting nano-bio interactions, and advancing the eco-responsible design and disposal of ENMs. The workshops led to 46 specific and ranked needs in research to ensure the safe development and use of ENMs (ICON 2008; 2010b).

More recently, the World Technology Evaluation Center, in collaboration with the NNI, published a comprehensive review of efforts to address nanotechnology EHS issues over the last 10 years with an eye toward the next 10 years (Nel et al. 2010). Written by leading researchers in the field, the report emphasizes the need for integrative and predictive capabilities to address a growing number of increasingly sophisticated nanomaterials. The report also discusses the potential for sophisticated approaches to address potential health risks posed by nanomaterials by providing the means to avoid harm rather than reacting to it—approaches that underpin green chemistry and sustainable development.

WHY ANOTHER STRATEGY IS NEEDED

Over the last 7 years, there has been considerable international effort to identify research needs for the development and safe use of nanotechnology products. Perhaps more than in the case of any other emerging technology, the possible consequences of new research and new developments have been analyzed and reanalyzed in advance of widespread commercialization. However, despite growing awareness of the challenges of developing nanotechnology-enabled products and using them safely, the connection and integration between generating data and analyses on emergent risks and developing strategies to prevent and manage them remain. Connection and integration will be necessary if the goal of informed oversight over an increasing array of products that depend on ENMs is to be achieved. Despite increasing budgets for nanotechnology-EHS research and a growing number of publications, regulators, decision-makers, and consumers still lack the information needed to make informed public health and environmental policy and regulatory decisions.

From reviewing the current state of research, and its relation to the needs of developers, regulators and users of ENMs, three particular points of "disconnect" are apparent. First, little research progress has been made in some fields, such as the effects of ingested ENMs on human health and the development of relevant and useful material-characterization techniques. Second, there is relatively little research on the potential health and environmental effects of more sophisticated ENMs, including active nanomaterials (materials that may change their biologic behavior in response to their environment or a signal), that are expected to enter commerce over the next decade. Third, system-integrative approaches that can address all forms of ENMs based on their properties and an understanding of the underlying biologic interactions that determine exposure and risk. Against that background, the identification of research needs in workshops over the last several years has been slow to be reflected in active-research pursuits. Common themes that have been identified in workshops include the need for standardized ENMs and harmonized methods for in vitro to in vivo validation in hazard assessments. Also evident are a number of new and rapidly evolving research approaches, including an increasing emphasis on high-throughput screening and predictive modeling, that are considered essential for managing the complexity of ENMs.[3]

That is not to say that progress is not being made—it clearly is. Repeated analyses of the safety challenges presented by ENMs have been accompanied by increases in both the quantity and the quality of research addressing these challenges. Assessing the impact of previous calls for targeted research is not straightforward—few (if any) authoritative studies have directly evaluated the response of funders, researchers, and practitioners to recommendations—but the evidence that does exist in the published literature suggests that progress has not been as strategically relevant as it could or should be. The documents listed in Table 1-1, suggest that expert communities have had a clear idea of what the key short-term questions are, but that there has been a failure to incorporate them

---

[3]Maynard et al. (2006) challenged the research community to develop high-throughput screening methods and predictive models for ENMs. In 2010, high-throughput screening and predictive modeling were central themes in a forward-looking evaluation of nanomaterial EHS challenges by Nel et al. (2010).

into integrated research strategies that lead to relevant and timely answers. The documents also indicate that the research community is involved in revisiting previously stated challenges rather than in demonstrating an awareness and appreciation of emerging challenges. For example, there is a repeated focus on simple nanomaterials, such as metal oxides, while researchers and developers are beginning to explore more complex and unconventional materials that are likely to require innovative means of understanding and addressing potential risks. Taken together, despite the substantial progress that has been made in recent years, there remains a need to develop an authoritative, integrative, and actionable research strategy that enables stakeholders in and outside the federal government to generate and apply new knowledge about the potential risks associated with ENMs in a timely and responsive manner. In particular, progress over the past decade and the current state of research suggest that there is a broader need for a new conceptual framework within which strategic EHS R&D can be planned, implemented, and reviewed.

SCOPE OF THIS REPORT

In response to the study request from EPA, the National Research Council (NRC) established the Committee to Develop a Research Strategy for Environmental, Health, and Safety Aspects of Engineered Nanomaterials. The committee was charged with developing and monitoring the implementation of an integrated research strategy to address the EHS aspects of ENMs. In response to the need for an authoritative, integrative, and actionable research strategy outlined above, this report will develop a conceptual framework for EHS-related research; establish a research plan with short- and long-term research priorities; and estimate resources necessary to implement this research plan. A subsequent report will evaluate research progress. The committee will consider current and emerging uses of ENMs and the scientific uncertainties related to physical and chemical properties, potential exposures, toxicity, toxicokinetics, and environmental fate of these materials. In its evaluation the committee will also consider existing research roadmaps and progress made in their implementation. A second report is to evaluate research progress and update the research priorities and resource estimates. The committee was not tasked with estimating the actual risk or benefits associated with EHS aspects of nanotechnology.

In addition to developing a conceptual framework for EHS-related research, this first report considers seven specific questions:

- What properties of ENMs need to be considered in assessing their potential exposures, toxicity, toxicokinetics, and environmental fate? What standard testing materials are needed?
- What methods and technologies are needed for detecting, measuring, analyzing, and monitoring ENMs? What gaps in analytic capability need to be addressed?

- What studies of ENM exposure, toxicology, toxicokinetics, human health effects, and environmental fate are needed for assessing the risks that they pose?
- What testing methods should be developed for assessing the potential toxicity, toxicokinetics, and environmental fate of ENMs?
- What models should be developed for predicting the effects of ENMs on human health and the environment?
- What are the research priorities for understanding life-cycle risks to humans and the environment from applications of nanotechnology?
- What criteria should be used to evaluate research progress?

ELEMENTS OF A NANOTECHNOLOGY ENVIRONMENTAL, HEALTH, AND SAFETY RESEARCH STRATEGY

In addressing its charge, the committee considered the key elements of a successful research strategy, as articulated in the *Review of the Federal Strategy for Nanotechnology-Related Environmental, Health, and Safety Research* (NRC 2009) (see Box 1-1). The 2009 NRC committee defined those elements and used them to evaluate the federal strategy. The committee based its discussion on the proposition that a strategy will address a set of defined goals, that there will be a plan to achieve the goals, and that metrics for success will indicate when the goals have been achieved. Because of the value of articulating and discussing the elements of a research strategy, the present committee has chosen to adopt the elements and build on them.

The elements of an effective research strategy were drawn from a historical perspective about how others had shaped their approaches to risk research and had considered the need to address a mix of both "exploratory" and "targeted" research. The elements are discussed below in the context of the current effort.

BOX 1-1 Elements of A Research Strategy

- Vision or statement of purpose
- Goals
- Evaluation of the existing state of the science
- Roadmap
- Evaluation
- Review
- Resources
- Mechanisms
- Accountability

Source: NRC 2009, p. 27.

Vision or Statement of Purpose

At the beginning of this chapter, the committee stated why a coherent risk-research strategy for nanomaterials is needed. Primary among the reasons is the rapid technologic advances that have resulted in the emergence of novel materials that have a potential for interaction with humans and ecosystems. There is considerable societal concern because of the potential, albeit unknown, risks. Scientific data and assessment are needed to determine the nature and extent of risks to the environment and health associated with a wide variety of existing and emerging nanomaterials. A strategy to address EHS issues would be a guide for scientists and decision-makers who need to set priorities for the use of limited resources while addressing the key risk-related questions.

Goals

Goals for the EHS risk-research strategy, articulated at the end of this chapter, are intended to guide the responsible development of novel nanomaterials and the management of existing nanomaterials and products to prevent and minimize their potential risks. The fulfillment of the goals will depend on the availability of resources and on the concerted efforts of government, academic, and industrial partnerships. Success in addressing these complex issues is possible only through interdisciplinary problem-solving.

Evaluation of the Existing State of the Science

A key component in the development of the risk-research strategy is the evaluation of the existing state of the science. Numerous efforts to catalog and evaluate relevant data and models have been made, and many have been published (see Table 1-1). Although it is not the intent of the current effort to evaluate that information de novo or exhaustively, a summary of the existing state of the science is provided in Chapter 3.

Roadmap

Given the complexity of the issues—the variety of the materials and applications of nanomaterial science not yet envisioned—it is critical that a roadmap (Chapters 5 and 6) be developed as a part of the current effort. The roadmap will need to address not only the path to short- and long-term research goals but the leveraging of available resources, institutions, and mechanisms both nationally and internationally. The roadmap will need to be flexible to incorporate new information and to be adjusted as knowledge and experience accumulate. It should also be responsive in the short term, providing approaches for environmental and human health protection even as the knowledge base is growing and the strategy is evolving.

Evaluation

A key element of an effective strategy is evaluation, to be discussed in Chapter 6. Measuring progress in research is inherently difficult. As discussed in NRC (2008), research cannot be evaluated on the basis of "ultimate outcomes," because outcomes often cannot be known or, in some cases, even expected. Instead, the 2008 report concludes that research should be evaluated on the basis of its quality, relevance, and effectiveness in addressing current priorities and future needs. The present report attempts to develop reasonable milestones of success along a defined timeline, considering the advice and perspective of NRC (2008).

Review

Like the committee that wrote NRC (2009), the present committee recognizes the need for periodic review and adjustment of the strategy as knowledge and experience accumulate. Review is critical in realizing the committee's vision of success, and processes for review will be discussed. It will require involvement of a broad array of stakeholders to ensure that a comprehensive perspective informs refinement of the strategy.

Resources

Resources to address the issues raised in the strategy will always be limited, but it is incumbent on this committee to estimate the resources needed to make reasonable progress toward success. It is beyond the scope of this effort to propose how such resources will be obtained or leveraged, but the magnitude of the resources should be considered if the nature of the problem is to be addressed.

Mechanisms

Optimal approaches and mechanisms for accomplishing exploratory and targeted research in the context of the strategy need to be discussed. As examples, the balance between government and industry funding and ways to enable interdisciplinary funding of collaborative research that crosses traditional agency or administrative boundaries will be considered.

Accountability

Accountability needs to be an element of the strategy. Who will "take ownership" of the overarching strategy? Who will assume or assign responsibility for individual aspects of the strategy? Who will be responsible for managing resources, ensuring review and stakeholder involvement, and developing the mechanisms discussed above? Those and other questions of accountability will be considered in Chapter 6.

PRIOR APPROACHES TO SETTING RESEARCH AGENDAS ON OTHER TOPICS

There are numerous examples of research agendas that have addressed major issues across a variety of domains. Spectacular successes of planned large-scale research and implementation strategies with defined objectives and end points are widely cited, including those of the Manhattan Project and the U.S. Human Genome Project. The Human Genome Project was implemented in 1991 with interrelated goals involving mapping of the human genome, the creation of a complete sequence of human DNA and the DNA of other organisms, and development of capabilities and technologies (for example, through the National Human Genome Research Institute). In the plan for the initial 5 years (1991-1995), cost estimates were made for sequencing the human genome. At the 5-year mark, a new plan for the next 5 years was elaborated (Collins et al. 1998). Progress toward the initial goals was charted, including analysis of quantifiable outcomes, and new goals were proposed. The initial working draft of the genome was published in 2000, ahead of schedule (Pennisi 2000a,b). The pace reflected technologic advances and the competition between the National Human Genome Research Institute and the Celera Corporation[4]. The project also benefited from a new paradigm of rapid and open data-sharing; sequence data were made available as they were generated.

The research agenda set by the National Research Council's Committee on Research Priorities for Airborne Particulate Matter (referred to as the PM Committee) is relevant to the charge of the present committee, although more narrowly defined in scope. The PM Committee was requested by Congress to address uncertainties in the scientific evidence related to airborne PM after the 1997 decision to establish a new National Ambient Air Quality Standard for fine PM. The uncertainties had been highlighted as the evidence on fine PM and health effects was reviewed. The PM Committee was charged with developing a multiyear research agenda, developing estimates of costs of implementing the strategy, monitoring progress, and evaluating the extent to which key uncertainties had been reduced. The PM Committee produced four reports related to its charge, the first in 1998 and the last in 2004 (NRC 1998, 1999, 2001, 2004).

The development of a framework for characterizing the sources of uncertainty was central to the PM Committee's approach (Figure 1-1). That toxicologic framework helped in identifying the major uncertainties and the com-

[4]Celera Corporation, a private company, worked in parallel with the government to sequence the human genome.

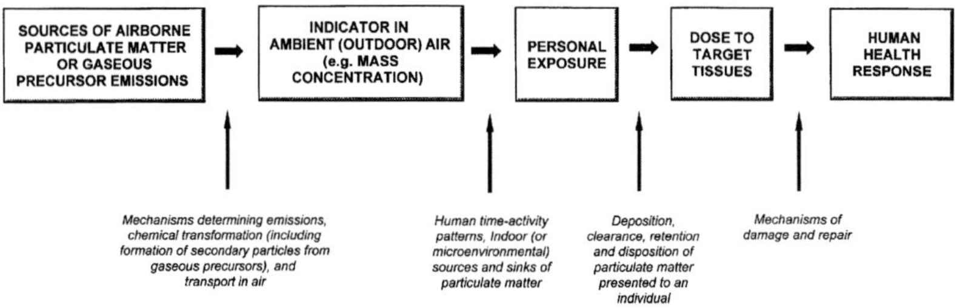

FIGURE 1-1 A general framework for integrating particulate-matter research. Source: NRC 1998, p. 35.

plementary research topics. Several of the research topics were overarching, such as the development and testing of air-quality models, and analysis and measurement. The committee proposed a "portfolio" of research to be undertaken over a 13-year span and estimated costs on a year-by-year basis.

In approaching its charge, the PM Committee elaborated on criteria for selecting its research topics, for characterizing progress on the agenda, and for evaluating the reduction of uncertainty. The committee selected three criteria for its initial list of research topics: scientific value, decision-making value, and feasibility and timing (NRC 1998). The first report (NRC 1998) provided definitions of the criteria and discussed their implementation. In its second and third reports (NRC 1999, 2001), the PM Committee added interaction, integration, and accessibility. With regard to the assessment of progress on the research agenda, the PM Committee screened existing approaches for useful models and took an evidence-based approach, involving surveying expenditures, research projects, and publications. For each topic, the final report covered the questions What has been learned? and What remains to be done? (NRC 2004).

The PM Committee reports provided useful examples for the present committee as it addressed its charge with regard to ENMs. The adoption of a toxicologic framework and the designation of research topics around the framework provided a needed transparent structure. The PM Committee's listing of operational criteria ensured clarity in how the research agenda was developed and tracked. (The present committee uses those criteria for evaluating research progress in Chapter 6.) Finally, the evidence-based approach ensured objectivity in the assessment of progress.

GOALS OF THIS STRATEGY

Despite the progress made to date, it remains imperative that we generate timely and relevant knowledge that will underpin the responsible development of technologies based on manipulating matter at the nanoscale. Nanoscale science and engineering are leading to remarkable new discoveries that have the potential to address major societal challenges while providing for substantial economic growth. Those discoveries are enabling the emergence of new technologies and the enhancement of existing technologies. Without strategic research on emergent risks associated with the new and enhanced technologies—and a clear understanding of how to prevent, manage, and avoid them—the future of safe and sustainable nanotechnology-based materials, products, and processes is uncertain. In today's fast-paced and interconnected world, a worthwhile economic and social return on government and industry investment in nanotechnology is unlikely to be fully realized without risk research, including research on translating knowledge into evidence-informed and socially responsible decision-making. Without the research, decision-makers will not have the tools and information that they need to develop safe and responsible technologies, trust in business and government to ensure safety could erode, and there will continue to be an inability to identify materials that pose important risks and to confidently differentiate them from materials and products that pose little or no risk.

There is a need to rethink and re-evaluate what is important from a human health and environmental perspective in addressing current and emerging ENMs and to establish a scientific foundation for risk-based decision-making. We need an EHS research strategy that is independent of any one stakeholder group, that reflects the interests of multiple types of stakeholders, that has as its primary aim protection of human and environmental health, that builds on past efforts and is flexible in anticipating and adjusting to emerging challenges, and that provides decision-makers and decision-influencers with timely, relevant, and accessible information.

Ten years after the establishment of the NNI, the emphasis of nanotechnology is shifting from research to commercialization. As it does, society cannot afford to remain entangled in confusion as the challenges and opportunities presented by nanotechnology are addressed. Although they were invaluable in their own right, previous attempts to identify research needs and develop research strategies have not brought needed clarity to nanotechnology EHS research needs, nor provided a relevant and actionable research strategy. The strategy presented here marks the development of a forward-looking, multiple stakeholder perspective that protects human and environmental health while reaping the potential benefits of nanoscale science.

In light of these needs, the goals of the research strategy are to generate scientific evidence that

- Guides approaches to environmental and human health protection even as our knowledge of ENMs is expanding and the research strategy itself is evolving.
- Makes it possible to identify and predict risks posed by nanomaterials with sufficient certainty to enable informed decisions on how the risks should be prevented, managed, or mitigated.

• Makes it possible to identify and evaluate the relative merits of various risk-management options, including measures to reduce the inherent hazard or exposure potential of nanomaterials.

Chapter 2 of this report presents a conceptual framework for considering the EHS risks associated with nanomaterials. Chapter 3 provides an overview of what is known about the EHS aspects of nanomaterials in the context of the conceptual framework and identifies knowledge gaps. Chapter 4 addresses cross-cutting tools needed to understand the relationship of ENM properties and their interactions with humans and the environment. Chapter 5 presents the committee's vision of the research agenda, including the recommended timing and costs of research activities. Chapter 6 describes implementation of the research strategy and how research progress will be evaluated.

REFERENCES

Aitken, R.J., S.M. Hankin, B. Ross, C.L. Tran, V. Stone, T.F. Fernandes, K. Donaldson, R. Duffin, Q. Chaudhry, T.A. Wilkins, S.A. Wilkins, L.S. Levy, S.A. Rocks, and A. Maynard. 2009. EMERGNANO: A Review of Completed and Near Completed Environment, Health and Safety Research on Nanomaterials and Nanotechnology. DEFRA (UK Department for Environment Food and Rural Affairs) Project CB0409. Report TM/09/01. Institute of Occupational Medicine, Edinburg, UK [online]. Available: http://www.safenano.org/Uploads/EMERGNANO_CB0409_Full.pdf [accessed Nov. 10, 2010].

Allianz/OECD (Allianz Center for Technology and Organisation for Economic Co-operation and Development). 2005. Small Sizes That Matter: Opportunities and Risks of Nanotechnologies: Report in Co-operation with the OECD International Futures Programme, C. Lauterwasser, ed. Allianz AG, München, and OECD International Futures Programme, Paris [online]. Available: http://www.oecd.org/data oecd/32/1/44108334.pdf [accessed Sept. 7, 2011].

Balbus, J.M., A.D. Maynard, V.L. Colvin, V. Castranova, G.P. Daston, R.A. Denison, K.L. Dreher, P.L. Goering, A.M. Goldberg, K.M. Kulinowski, N.A. Monteiro-Riviere, G. Oberdörster, G.S. Omenn, K.E. Pinkerton, K.S. Ramos, K.M. Rest, J.B. Sass, E.K. Silbergeld, and B.A. Wong. 2007. Meeting report: Hazard assessment for nanoparticles-report from an interdisciplinary workshop. Environ. Health Perspect. 115(11):1654-1659.

Bradley, J. 2010. Nanotech's Evolving Environmental Health, and Safety Landscape. Presentation at Nanosafe 2010: International Conference on Safe Production and Use of Nanomaterials, Nov. 16-18, 2010, Minatec, France [online]. Available: http://www.nanosafe.org/home/liblocal/docs/Nanosafe%202010/2010_oral%20presentations/PL0a_Bradley.pdf [accessed Dec. 17, 2010].

Breggin, L., R. Falkner, N. Jaspers, J. Pendergrass, and R. Porter. 2009. Securing the Promise of Nanotechnologies: Towards Transatlantic Regulatory Cooperation. London: Chatham House [online]. Available: http://www.chathamhouse.org.uk/files/14692_r0909_nanotechnologies.pdf [accessed Apr. 14, 2011].

Collins , F.S., A. Patrinos, E. Jordan, A. Chakravarti, R. Gesteland, L. Walters, and the members of the DOE and NIH planning groups. 1998. New goals for the U.S. Human Genome Project: 1998-2003. Science 282(5389): 682-689.

Council of Canadian Academies. 2008. Small is Different: A Science Perspective on the Regulatory Challenges of the Nanoscale, Report of the Expert Panel of Nanotechnology. Ottawa: Council of Canadian Academies [online]. Available: http://www.scienceadvice.ca/uploads/eng/assessments%20and%20publications%20and%20news%20releases/nano/(2008_07_10)_report_on_nanotechnology.pdf [accessed Apr. 13, 2011].

Davis, J.M., and V.M. Thomas. 2006. Systematic approach to evaluating trade-offs among fuel options: The lessons of MTBE. Ann. N.Y. Acad. Sci. 1076:498-515.

DEFRA (Department for Environment, Food and Rural Affairs). 2005. Characterising the Potential Risks Posed by Engineered Nanoparticles: A First UK Government Research Report. Department for Environment, Food and Rural Affairs, Her Majesty's Government, London [online]. Available: http://www.nanowerk.com/nanotechnology/reports/reportpdf/report12.pdf [accessed Apr. 13, 2011].

DEFRA (Department for Environment, Food and Rural Affairs). 2007. Characterising the Potential Risks Posed by Engineered Nanoparticles: A Second UK Government Research Report. Department for Environment, Food and Rural Affairs, Her Majesty's Government, London [online]. Available: http://archive.defra.gov.uk/environment/quality/nanotech/documents/nanoparticles-riskreport07.pdf [accessed Apr. 13, 2011].

Denison, R.A. 2005. A Proposal to Increase Federal Funding of Nanotechnology Risk Research to at Least $100 Million Annually. Environmental Defense Fund. April 2005 [online]. Available: http://www.edf.org/documents/4442_100milquestionl.pdf [accessed Apr. 13, 2011].

EC (European Commission). 2005. Communication from the Commission to the Council, the European Parliament and the Economic Social Committee: Nanosciences and Nanotechnologies: An Action Plan for Europe 2005-2009. European Commission, Brussels [online]. Available: http://ec.europa.eu/research/industrial_technologies/pdf/nano_action_plan_en.pdf [accessed Nov. 9, 2010].

EDF/DuPont (Environmental Defense Fund and DuPont). 2007. Nano Risk Framework. Environmental Defense Fund, Washington, DC, and DuPont, Wilmington, DE. June 2007 [online]. Available: http://nanoriskframework.com/page.cfm?tagID=1081 [accessed Mar. 17, 2010].

EFSA (European Food Safety Authority). 2009. Scientific Opinion: The Potential Risks Arising from Nanoscience and Nanotechnologies on Food and Feed Safety. EFSA J. 958:1-39 [online]. Available: http://www.efsa.europa.eu/fr/scdocs/doc/sc_op_ej958_nano_en,0.pdf [accessed Apr. 14, 2011].

EPA (U.S. Environmental Protection Agency). 2005. Nanotechnology White Paper. External Review Draft. Nanotechnology Workgroup, Science Policy Council, U.S. Environmental Protection Agency, Washington, DC. December 2, 2005 [online]. Available: http://www.epa.gov/osa/pdfs/EPA_nanotechnology_white_paper_external_review_draft_12-02-2005.pdf [accessed Nov. 10, 2010].

EPA (U.S. Environmental Protection Agency). 2007. Nanotechnology White Paper. EPA 100/B-07/001. Office of the Science Advisor, Science Policy Council, U.S. Environmental Protection Agency, Washington, DC. February 2007 [online]. Available: http://www.epa.gov/osainter/pdfs/nanotech/epa-nanotechnology-whitepaper-0207.pdf [accessed Apr. 13, 2011].

EPA (U.S. Environmental Protection Agency). 2009. Nanomaterial Research Strategy. EPA 620/K-09/011. Office of Research and Development, U.S. Environmental Protection Agency, Washington, DC. June 2009 [online]. Available: http://www.epa.gov/nano science/files/nanotech_research_strategy_final.pdf [accessed Nov. 10, 2010].

ETC Group (Action Group of Erosion, Technology and Concentration). 2003. Size Matters! No Small Matter II: The Case for a Global Moratorium. Occasional Paper 7(1). ETC Group, Ottawa, Canada [online]. Available: http://www.etcgroup.org/ upload/publication/165/01/occ.paper_nanosafety.pdf [accessed Nov. 10, 2010].

FAO/WHO (Food and Agriculture Organization of the United Nation and World Health Organization) 2009. FAO/WHO Expert Meeting on the Application of Nanotechnologies in the Food and Agriculture Sectors: Potential Food Safety Implications. Meeting Report. Food and Agriculture Organization of the United Nation, Rome, Italy, and World Health Organization, Geneva, Switzerland [online]. Available: http://www.fao.org/ag/agn/agns/files/FAO_WHO_Nano_Expert_Meeting_Report_ Final.pdf [accessed Apr. 14, 2011].

FDA (U.S. Food and Drug Administration). 2007. Nanotechnology. A Report of the U.S. Food and Drug Administration Nanotechnology Task Force. July 25, 2007. U.S. Department of Health and Human Services, Public Health Service, Food and Drug Administration, Rockville, MD [online]. Available: http://www.fda.gov/downloads/ ScienceResearch/SpecialTopics/Nanotechnology/ucm110856.pdf [accessed Apr. 13, 2011].

Ford, E. 2005. Recommendations for Nanotechnology ESH Research. Presentation at the American Institute of Chemical Engineers (AIChE) Annual Meeting, November 3, 2005, Cincinnati, OH [online]. Available: http://www.chemicalvision2020.org/ pdfs/nano_recs.pdf [accessed Nov. 10, 2010].

GAO (U.S. Government Accountability Office). 2008. Nanotechnology: Better Guidance is Needed to Ensure Accurate Reporting of Federal Research Focused on Environmental, Health and Safety Risks. GAO-08-402. Washington, DC: U.S. Government Accountability Office [online]. Available: http://www.gao.gov/new.items/ d08402.pdf [accessed Apr. 14, 2011].

GAO (U.S. Government Accountability Office). 2010. Nanotechnology: Nanomaterials Are Widely Used in Commerce, but EPA Faces Challenges in Regulating Risk. GAO-10-549. Washington, DC: U.S. Government Accountability Office [online]. Available: http://www.gao.gov/new.items/d10549.pdf [accessed Apr. 14, 2011].

HM Government (Her Majesty's Government). 2010. UK Nanotechnologies Strategy: Small Technologies, Great Opportunities. Her Majesty's Government, London. March 2010 [online]. Available: http://www.bis.gov.uk/assets/BISPartners/GoScie nce/Docs/U/10-825-uk-nanotechnologies-strategy [accessed Apr. 14, 2011].

Höck, J. 2008. Proceedings of the Workshop on Research Projects on the Safety of Nanomaterials: Reviewing the Knowledge Gaps, 17-18 April 2008, Brussels. DG Research, European Commission [online]. Available: ftp://ftp.cordis.europa.eu/pu b/nanotechnology/docs/final_report.pdf [accessed Apr. 13, 2011].

Hwang, D., and J. Bradley. 2010. The Recession's Ripple Effect on Nanotech. Small Times, August 12, 2010 [online]. Available: http://www.electroiq.com/index/display/nano tech-article-display/5049629279/articles/small-times/nanotechmems/materials/gener al/2010/april/the-recession_s_ripple.html [accessed Apr. 13, 2011].

ICFI (ICF International). 2006. Characterizing the Environmental, Health and Safety Implications of Nanotechnology: Where Should the Federal Government Go From Here? ICF International, Fairfax, VA.

ICON (International Council on Nanotechnology). 2008. Towards Predicting Nano-Biointeractions: An International Assessment of Nanotechnology Environment, Health and Safety Research Needs. No. 4. International Council on Nanotechnology, Rice University, Houston, TX. May 1, 2008 [online]. Available: http://cohesion.rice.edu/centersandinst/icon/emplibrary/icon_rna_report_full2.pdf [accessed Feb. 16, 2011].

ICON (International Council on Nanotechnology). 2010a. International Assessment of Research Needs for Nanotechnology Environmental, Health, and Safety. International Council on Nanotechnology, Rice University, Houston, TX [online]. Available: http://icon.rice.edu/projects.cfm?doc_id=12973 [accessed Nov. 22, 2010].

ICON (International Council on Nanotechnology). 2010b. Advancing the Eco-Responsible Design and Disposal of Engineered Nanomaterials: An International Workshop. No. 5. International Council on Nanotechnology, Rice University, Houston, TX. February 2010 [online]. Available: http://cohesion.rice.edu/centersandinst/icon/emplibrary/ICON_Eco-Responsible_Design_and_Disposal%20of_Engineered_Nanomaterials_Full_Report.pdf [accessed Nov. 10, 2010].

Joy, B. 2000. Why the Future Doesn't Need Us. Wired Magazine, August 4, 2000 [online]. Available: http://www.wired.com/wired/archive/8.04/joy.html?pg=1&topic=&topic_set= [accessed Nov. 10, 2010].

Kaluza, S., J.K. Balderhaar, B. Orthen, B. Honnert, E. Jankowska, P. Pietrowski, M.G. Rosell, C. Tanarro, J. Tejedor, and A. Zugasti. 2009. Literature Review: Workplace Exposure to Nanoparticles, J. Kosk-Bienko, ed. European Agency for Safety and Health at Work [online]. Available: http://osha.europa.eu/en/publiccations/literature_reviews/workplace_exposure_to_nanoparticles [accessed Apr. 14, 2011].

Lloyd's. 2007. Nanotechnology Recent Developments, Risks and Opportunities. Lloyd's Emerging Risks Team Report [online]. Available: http://www.lloyds.com/~/media/c6f557f00581437ea4b9325293731275.ashx [accessed Sept. 7, 2011].

Luther, W., ed. 2004. Technological Analysis: Industrial Application of Nanomaterials-Chances and Risks. Future Technologies No. 54. Future Technologies Division of VDI Technologiezentrum GmbH, Düsseldorf. August 2004 [online]. Available: http://www.zukuenftigetechnologien.de/11.pdf [accessed Apr. 13, 2011].

Lux Research. 2008a. Nanomaterials State of the Market Q3 2008: Stealth Success, Broad Impact. Lux Research, July 1, 2008 [online]. Available: https://portal.luxresearchinc.com/research/document_excerpt/3735 [accessed Dec. 17, 2010].

Lux Research. 2008b. Overhyped Technology Starts to Reach Potential: Nanotech to Impact $3.1 Trillion in Manufactured Goods in 2015. Press release: July 22, 2008 [online]. Available: http://www.luxresearchinc.com/press/RELEASE_Nano-SMR_7_22_08.pdf [accessed Dec. 17, 2010].

Lux Research. 2009a. The Wizards of Nanointermediates: Assessing Catalysts, Coatings, and Composites on the Lux Innovation Grid. Lux Research, September 2, 2009 [online]. Available: https://portal.luxresearchinc.com/research/document_excerpt/5383 [accessed Dec. 17, 2010].

Lux Research. 2009b. The Recession's Ripple Effect on Nanotech, State of the Market Research. Lux Research, June 9, 2009 [online]. Available: https://portal.luxresearchinc.com/research/document_excerpt/4995 [accessed Feb. 22, 2010].

Maynard, A.D. 2006. Nanotechnology: A Research Strategy for Addressing Risk. Project on Emerging Nanotechnologies PEN 3. Washington, DC: Woodrow Wilson International Center for Scholars. July 2006 [online]. Available: http://www.nanotechproject.org/process/assets/files/2707/77_pen3_risk.pdf [accessed Nov. 10, 2010].

Maynard, A.D., R.J. Aitken, T. Butz, V. Colvin, K. Donaldson, G. Oberdörster, M.A. Philbert, J. Ryan, A. Seaton, V. Stone, S.S. Tinkle, L. Tran, N.J. Walker, and D.B. Warheit. 2006. Safe handling of nanotechnology. Nature 444(7117):267-269.

Maynard, A.D. 2010. Rethinking Nanotechnology—Responding to a Request for Information on the U.S. Nanotechnology Strategic Plan. 2020 Science, August 30, 2010 [online]. Available: http://2020science.org/2010/08/30/rethinking-nanotechnology-responding-to-a-request-for-information-on-the-us-nanotechnology-strategic-plan/ [accessed November 28, 2011].

NEHI (Nanotechnology Environmental Health Implications Working Group). 2006. Environmental, Health, and Safety Research Needs for Engineering Nanoscale Materials. Nanotechnology Environmental Health Implications Working Group, Nanoscale Science, Engineering, and Technology Subcommittee, Committee on Technology, National Science and Technology Council. September 2006 [online]. Available: http://www.nano.gov/NNI_EHS_research_needs.pdf [accessed July 12. 2011].

NEHI (Nanotechnology Environmental Health Implications Working Group). 2007. Prioritization of Environmental, Safety and Health Research Needs for Engineered Nanoscale Materials: An Interim Document for Public Comment. Nanotechnology Environmental Health Implications Working Group, Nanoscale Science, Engineering, and Technology Subcommittee, Committee on Technology, National Science and Technology Council. August 2007 [online]. Available: http://nanotech.law.asu.edu/Documents/2010/08/Prioritization_EHS_Research_Needs_Engineered_Nanoscale_Materi als_527_8119.pdf [accessed July 13, 2011].

NEHI (Nanotechnology Environmental Health Implications Working Group). 2008. Strategy for Nanotechnology-Related Environmental, Health, and Safety Research. National Nanotechnology Initiative. Nanotechnology Environmental Health Implications Working Group, Subcommittee on Nanoscale Science, Engineering, and Technology, Committee on Technology, National Science and Technology Council. February 2008 [online]. Available: http://www.nano.gov/NNI_EHS_Resear ch_Strategy.pdf [accessed July 12. 2011].

NEHI (Nanotechnology Environmental Health Implications Working Group). 2010. National Nanotechnology Initiative 2011 Environmental, Health, and Safety Strategy - Draft for Public Comment, December 2010. Nanotechnology Environmental Health Implications Working Group, Subcommittee on Nanoscale Science, Engineering, and Technology, Committee on Technology, National Science and Technology Council [online]. Available: http://ethics.iit.edu/NanoBank/Draft-2011-NNI-EHS-Research-Strategy.pdf [accessed July 12. 2011].

NEHI (Nanotechnology Environmental Health Implications Working Group). 2011. National Nanotechnology Initiative 2011 Environmental, Health, and Safety Strategy, October 2011. Nanotechnology Environmental Health Implications Working Group, Subcommittee on Nanoscale Science, Engineering, and Technology, Committee on Technology, National Science and Technology Council [online]. Available: http://www.nano.gov/sites/default/files/pub_resource/nni_2011_ehs_research_strategy.pdf [accessed Nov. 23, 2011].

Nel, A., D. Grainger, P. Alvarez, S. Badesha, V. Castranova, M. Ferrari, H. Godwin, P. Grodzinski, J. Morris, N. Savage, N. Scott, and M. Wiesner. 2010. Nanotechnology environmental, health and safety issues. Pp. 107-156 in WTEC Panel Report on Nanotechnology Research Directions for Societal Needs in 2020: Retrospective and Outlook, M.C. Roco, C.A. Mirkin, and M.C. Hersam, eds. Springer [online].

Available: http://www.wtec.org/nano2/Nanotechnology_Research_Directions_to_2020/Nano_Resarch_Directions_to_2020.pdf [accessed Feb. 16, 2011].

NIOSH (National Institute for Occupational Safety and Health). 2005. Strategic Plan for NIOSH Nanotechnology Research, Filling the Knowledge Gaps, Draft [online]. Available: http://www.cdc.gov/niosh/topics/nanotech/strat_plan.html [accessed Aug. 1, 2011].

NIOSH (National Institute for Occupational Safety and Health). 2008. Strategic Plan for NIOSH Nanotechnology Research and Guidance: Filling the Knowledge Gaps. Draft Report. National Institute for Occupational Safety and Health, Centers for Disease Control and Prevention, U.S. Department of Health and Human Services. February 26, 2008 [online]. Available: http://www.cdc.gov/niosh/topics/nanotech/pdfs/NIOSH_Nanotech_Strategic_Plan.pdf [accessed Apr. 14, 2011].

NIOSH (National Institute for Occupational Safety and Health). 2010. Strategic Plan for NIOSH Nanotechnology Research and Guidance: Filling the Knowledge Gaps. National Institute for Occupational Safety and Health, Centers for Disease Control and Prevention, U.S. Department of Health and Human Services [online]. Available: http://www.cdc.gov/niosh/docs/2010-105/pdfs/2010-105.pdf [accessed Nov. 10, 2010].

NNI-ChI CBAN ESH Working Group (National Nanotechnology Initiative-Chemical Industry Consultative Board for Advancing Nanotechnology, Environmental Safety and Health Working Group). 2007. Recommended Topics for R&D Activities: Toxicological Hazard Assessment of Nanomaterials. The Chemical Industry Vision2020 Technology Partnership (Vision 2020) [online]. Available: http://www.chemicalvision2020.org/pdfs/cban_recommendedtopics.pdf [accessed Nov. 10, 2010].

NRC (National Research Council). 1998. Research Priorities for Airborne Particulate Matter. I. Immediate Priorities and Long-Range Research Portfolio. Washington, DC: National Academy Press.

NRC (National Research Council). 1999. Research Priorities for Airborne Particulate Matter. II. Evaluating Research Progress and Updating the Portfolio. Washington, DC: National Academy Press.

NRC (National Research Council). 2001. Research Priorities for Airborne Particulate Matter. III. Early Research Progress. Washington, DC: National Academy Press.

NRC (National Research Council). 2004. Research Priorities for Airborne Particulate Matter. IV. Continuing Research Progress. Washington, DC: National Academies Press.

NRC (National Research Council). 2008. Evaluating Research Efficiency in the U.S. Environmental Protection Agency. Washington, DC: National Academies Press.

NRC (National Research Council). 2009. Review of the Federal Strategy for Nanotechnology-Related Environmental, Health, and Safety Research. Washington, DC: National Academies Press.

NSET (Nanoscale Science, Engineering, and Technology Subcommittee). 2004. The National Nanotechnology Initiative Strategic Plan. Subcommittee on Nanoscale Science, Engineering, and Technology, Committee on Technology, National Science and Technology Council. December 2004 [online]. Available: http://neutrons.ornl.gov/workshops/nni_05/nni_strategic_plan_0412.pdf [accessed July 12, 2011].

NSET (Nanoscale Science, Engineering, and Technology Subcommittee). 2006. The National Nanotechnology Initiative: Research and Development Leading to a Revolution in Technology and Industry: Supplement to the President's FY 2007 Budget. Subcommittee on Nanoscale Science, Engineering, and Technology, National Sci-

ence and Technology Council. July 2006 [online]. Available: http://jcots.state.va.us/2007%20Content/Materials/2007%20NNI%20Budget.pdf [accessed July 12, 2011].
NSET (Nanoscale Science, Engineering, and Technology Subcommittee). 2010. The National Nanotechnology Initiative: Research and Development Leading to a Revolution in Technology and Industry: Supplement to the President's FY 2011 Budget. Subcommittee on Nanoscale Science, Engineering, and Technology, National Science and Technology Council. February 2010 [online]. Available: https://engineering.purdue.edu/nanotrees/files/NNI_2011_budget_supplement.pdf [accessed July 13, 2011].
Oberdörster, G., V. Stone, and K. Donaldson. 2007. Toxicology of nanoparticles: A historical perspective. Nanotoxicology. 1(1):2-25.
PCAST (President's Council of Advisors on Science and Technology). 2010. Report to the President and Congress on the Third Assessment of the National Nanotechnology Initiative. March 2010 [online]. Available: http://www.whitehouse.gov/sites/default/files/microsites/ostp/pcast-nano-report.pdf [accessed Nov. 9, 2010].
Pennisi, E. 2000a. Chromosone 21 done. Phase two begun. Science. 288(5468):939.
Pennisi, E. 2000b. And the gene number is…? Science. 288(5469):1146-1147.
RCEP (Royal Commission on Environmental Pollution). 2008. Twenty-Seventh Report: Novel Materials in the Environment: The Case of Nanotechnology. Royal Commission on Environmental Pollution. November 2008 [online]. Available: http://www.official-documents.gov.uk/document/cm74/7468/7468.pdf [accessed Nov. 10, 2010].
Renn, O., and M. Roco. 2006. White Paper on Nanotechnology Risk Governance. White Paper No. 2. International Risk Governance Council, Geneva [online]. Available: http://www.irgc.org/IMG/pdf/IRGC_white_paper_2_PDF_final_version-2.pdf [accessed Apr. 13, 2011].
RS/RAE (The Royal Society and The Royal Academy of Engineering). 2004. Nanoscience and Nanotechnologies: Opportunities and Uncertainties. London: The Royal Society. July 2004 [online]. Available: http://www.nanotec.org.uk/report/Nano%20report%202004%20fin.pdf [accessed Nov. 9, 2010].
SCCP (Scientific Committee on Consumer Products). 2007. Opinion on Safety of Nanomaterials in Cosmetic Products. SCCP/1147/07. Scientific Committee on Consumer Products, Health and Consumer Protection Directorate, European Commission. December 2007 [online]. Available: http://ec.europa.eu/health/ph_risk/committees/04_sccp/docs/sccp_o_123.pdf [accessed Apr. 13, 2011].
SCENIHR (Scientific Community on Emerging and Newly Identified Health Risks). 2006. Opinion on the Appropriateness of Existing Methodologies to Assess the Potential Risks Associated with Engineered and Adventitious Products of Nanotechnologies. SCENIHR 002/05. Scientific Community on Emerging and Newly Identified Health Risks, Health & Consumer Protection Directorate-General, European Commission. March 2006 [online]. Available: http://www.oecd.org/dataoecd/52/24/37999466.pdf [accessed Nov. 9, 2010].
SCENIHR (Scientific Community on Emerging and Newly Identified Health Risks). 2007. Opinion on the Appropriateness of the Risk Assessment Methodology in Accordance with the Technical Guidance Documents for New and Existing Substances for Assessing the Risks of Nanomaterials. Scientific Community on Emerging and Newly Identified Health Risks, Health & Consumer Protection Directorate-General, European Commission. June 2007 [online]. Available: http://ec.europa.eu/health/ph_risk/committees/04_scenihr/docs/scenihr_o_010.pdf [accessed Apr. 13, 2011].
SCENIHR (Scientific Committee on Emerging and Newly Identified Health Risks). 2009. Risk Assessment of Products of Nanotechnologies. Scientific Committee on Emerg-

ing and Newly Identified Health Risks, Health & Consumer Protection Directorate-General, European Commission. January 2009 [online]. Available: http://ec.europa.eu/health/ph_risk/committees/04_scenihr/docs/scenihr_o_023.pdf [accessed Apr. 14, 2011].

Schierow, L.J. 2008. Engineered Nanoscale Materials and Derivate Products: Regulatory Challenges. CRC Report for Congress RL34332. Congressional Research Service, Washington, DC. January 22, 2008 [online]. Available: http://www.fas.org/sgp/crs/misc/RL34332.pdf [accessed Apr. 14, 2011].

Sibley, L. 2009. Cleantech Sector Offers Big Potential for Nanomaterials. Cleantech Group LLC, September 22, 2009 [online]. Available: http://www.cleantech.com/news/5044/energy-applications-hold-big-potent [accessed Dec. 17, 2010].

Stone, V., S. Hankin, R.J. Aitken, K. Aschberger, A. Baun, F. Christensen, T. Fernandes, S.F. Hansen, N.B. Hartmann, G. Hutchison, H. Johnson, C. Micheletti, S. Peters, B. Ross, B. Sokull- Kluettgen, D. Stark, and L. Tran. 2010. Engineered Nanoparticles: Review of Health and Environmental Safety (ENRHES). Final Report. Edinburgh Napier University (ENU), the Institute of Occupational Medicine(IOM), Edinburg, UK, the Technical University of Denmark (DTU), the Institute for Health and Consumer Protection of the European Commission's Joint Research Centre (JRC), and Institute of Nanotechnology (IoN). European Commission [online]. Available: http://www.nanowerk.com/nanotechnology/reports/reportpdf/report133.pdf [accessed Dec. 27, 2011].

Swiss Re (Swiss Reinsurance Company). 2004. Nanotechnology: Small Matter, Many Unknowns. Swiss Reinsurance Company, Zurich [online]. Available: http://www.nanowerk.com/nanotechnology/reports/reportpdf/report93.pdf [accessed Nov. 9, 2010].

Teague, E.C. 2005. Testimony on Environmental and Safety Impacts of Nanotechnology before Committee on Science and Technology, 109th Cong., November 17, 2005 [online]. Available: http://ftp.resource.org/gpo.gov/hearings/109h/24464.pdf [accessed March 1, 2011].

The White House. 2000. President Clinton's Address to Caltech on Science and Technology, California Institute of Technology, Pasadena, CA. January 21, 2000 [online]. Available: http://www.dtrends.com/Nanotech/nano_clinton.html [accessed Feb. 16, 2011].

UKHL (United Kingdom House of Lords). 2010. Nanotechnologies and Food. HL Paper 22-1. United Kingdom House of Lords, Science and Technology Committee. London: The Stationery Office. January 2010 [online]. Available: http://www.publications.parliament.uk/pa/ld200910/ldselect/ldsctech/22/22i.pdf [accessed Nov. 10, 2010].

VCI (Verband der Chemischen Industrie e.V. German Chemical Industry Association). 2008. Responsible Production and Use of Nanomaterials. German Chemical Industry Association. March 11, 2008 [online]. Available: http://www.vci.de/template_downloads/tmp_VCIInternet/122306Nano_Responsible_Production.pdf?DokNr=122306&p=101 [accessed Apr. 14, 2011].

Vision 2020/SRC (Chemical Industry Vision 2020 Technology Partnership and Semiconductor Research Corporation). 2005. Joint NNI-ChI CBAN and SRC CWG5 Nanotechnology Research Needs Recommendations [online]. Available: http://www.chemicalvision2020.org/pdfs/chem-semi%20ESH%20recommendations.pdf [accessed Nov. 10, 2010].

Wray, P. 2010. Nano Products Taking a Hit During the Current Recession CTT, March 12, 2010 [online]. Available: http://ceramics.org/ceramictechtoday/2010/03/12/nano-products-taking-a-hit-in-hard-times/ [accessed Nov. 23, 2011].

2

A Conceptual Framework for Considering Environmental, Health, and Safety Risks of Nanomaterials

THE NATURE OF THE CHALLENGE

The rapid emergence of engineered nanomaterials (ENMs) and their use in diverse products imply their eventual and inevitable appearance in the biosphere. As discussed in Chapter 1, the environmental and human health risks posed by these novel materials remain largely unknown, but the materials' widespread use provides a strong motivation for investment in research directed at potential adverse effects. The vast variety of nanomaterials and their novel properties provide a strong basis for systematic, coordinated, and integrated research efforts to understand what properties of the materials influence their hazard and exposure potential and what applications present the greatest likelihood of exposure and adverse effects on human health and the environment.

ENMs are a subset of the broader field of nanotechnology, which is defined by the National Nanotechnology Initiative (NNI) as "the understanding and control of matter at dimensions between approximately 1 and 100 nanometers, where unique phenomena enable novel applications. Encompassing nanoscale science, engineering, and technology, nanotechnology involves imaging, measuring, modeling, and manipulating matter at this length scale" (NSET 2010a).

Scale-specific properties and phenomena are at the heart of current interest and investment in ENMs. A substance can be designed and engineered at the nanoscale to behave in a particular and useful way, thereby potentially adding value to an existing product or becoming the basis of a completely new product. Scale-specific properties of nanomaterials expand the possibilities for making new products. But the same scale-specific properties are at the center of concerns about possible new risks: if a new material behaves in novel ways, what are the chances that this behavior will lead to harm to people and the environment?

The multiplicity of ENM variants makes material-by-material assessment impractical. That heterogeneity in nanomaterials, characterized by distributions of properties, has spurred efforts to generalize about exposure and hazard potential in relation to these properties, rather than considering risks for specific types of materials. Initial attempts point to complexities in understanding risks of ENMs (Dreher 2004). For example, the size range used to describe ENMs—1-100 nm—has relatively little bearing itself in determining the risk to people or the environment (see, for example, Auffan et al. 2009; Drezek and Tour 2010). Risk "problems" associated with ENMs have been formulated in terms of established "technologic" characteristics of ENMs (such as particle size) that do not appropriately reflect the potential for harm.

Framing risks associated with an ENM in terms of established definitions provides some insight into emergent risks. For example, exposure potential may be enhanced as particle size decreases to the point where novel physicochemical properties begin to dominate behavior. At the same time, a focus on particle size may highlight issues that are not relevant while shifting attention from such properties as reactivity that may be more relevant to determining risks (for example, Maynard 2011; Maynard et al. 2011a). Consequently, there is substantial uncertainty in understanding of the risks associated with the products of nanotechnology, leading to confusion on prioritizing, and addressing these risks—a confusion that is illustrated in many reports on risk. (See discussion in Chapter 1.)

In making risk-based decisions—whether translating an innovative idea into a new product, crafting new regulations, or developing a risk-research strategy—effective problem formulation is essential (NRC 2009). Formulating the environmental, health, and safety (EHS) "problems" presented by ENMs has proved challenging, as documented by research efforts over the last decade.

DEVELOPING A STRATEGY AND A CONCEPTUAL FRAMEWORK

In addressing the challenges presented by ENMs, the committee notes that there is a distinction between a *research strategy* and a *research agenda*. The committee has developed a strategy that provides a principle-based approach to sustaining an agenda for EHS research that will be accountable and adaptive as ENMs change, diversify, and expand in use. In this chapter, the committee describes the research framework for its strategy; later chapters identify data gaps to be addressed by the research strategy. The generation of findings for risk assessment is considered here as an evolving process based on the integration of various research efforts rather than as a static "deliverable." There will be an ongoing need to inform decision-making in advance of product development and to consider uncertainty coming from incomplete information on future production quantities, ENM properties, and uses of nanomaterials. An evolving and iterative process provides feedback for adjusting research priorities and provides the information needed to implement risk-management strategies aimed at reducing the potential for harm to human health and the environment. That feedback also informs the design, manufacture, and use of future ENMs.

The conceptual framework, described later in this chapter, reflects a coordinated, strategic research effort that is characterized by three key features:

- A reliance on principles that help to identify emergent, plausible, and severe risks resulting from designing and engineering materials at the nanoscale, rather than an adherence to rigid definitions of ENMs.
- A value-chain and life-cycle perspective that considers the potential harm originating in the production and use of nanomaterials, nanomaterial-containing products, and the wastes generated.
- A focus on determining how nanomaterial properties affect key biologic processes that are relevant to predicting both hazard and exposure; for example, nanomaterial-macromolecular interactions that govern processes ranging from protein folding (a basis for toxicity) to the adsorption of humic substances (that may influence mobility or bioavailability of the materials).

Environmental and human health risk assessment of nanomaterials is severely limited by lack of information on exposure to these materials (for example, information on fate, transport, and transformations) and on the hazards that they present. In contrast with previous research strategies that took a sequential approach to evaluating exposure and hazard for assessing nanomaterial-related risks, the committee's framework considers evaluations of hazards and exposure as processes that occur in tandem, and it accounts for the wide variety of matrices and transformations of nanomaterials along the value chain and across the life cycle (discussed in more detail later in this chapter).

The framework is to be implemented through a research agenda that begins with understanding how nanomaterial properties may affect fundamental processes—processes that are common in determining both exposures and hazards. By focusing on these processes, the goal of advancing exposure and hazard assessment under conditions of uncertainty can be addressed in a predictive and generalizable fashion that helps to inform decision-making on current and future nanomaterials. Knowledge of these processes has immediate applicability in comparing risks among materials and providing criteria for establishing priorities for research on nanomaterials that are on the market, for providing feedback on research needs and priorities, and for providing evidence needed to reduce the risks posed by nanomaterials that are on the market or are under development.

The sections below address the utility of risk assessment in framing a research strategy for the EHS aspects of nanomaterials, the conceptual framework that is informed by risk assessment, and the principles for setting priorities among research needs on the basis of the properties of nanomaterials.

RISK-ASSESSMENT CONSIDERATIONS REGARDING NANOMATERIALS

In developing this chapter, the committee found useful guidance in *Science and Decisions: Advancing Risk Assessment* (NRC 2009), which offers recommendations for addressing risks in the modern world. The report examines near-term (2-5 y) and longer-term (10-20 y) solutions focusing on human health risk assessment, but it also considers the implications for ecologic risk assessment. The report focused on two broad goals in its evaluation: improving the technical analysis that supports risk assessment and improving the utility of risk assessment. Although that committee concluded that technical improvements are necessary, it suggested retaining the four basic elements of risk assessment—hazard identification, exposure assessment, dose-response assessment, and risk characterization—originally articulated in *Risk Assessment in the Federal Government: Managing the Process* (NRC 1983). Technical improvements are needed in approaches to uncertainty and variability analysis and in dose-response analysis. With regard to improving the utility of risk assessment, the committee authoring that report focused on improvements in scoping the problem at hand and understanding a broad set of risk-management options so that the ensuing risk assessment would be more relevant to the questions that decision-makers might ask of the scientific-knowledge base. An important conclusion of the committee's work was that risk assessment, rather than being viewed as an end in itself, should be considered as a method for informing research and commercialization efforts and for evaluating the relative merits of various risk-management options.

In the context of the development of an EHS risk-research strategy for ENMs, NRC (2009) has much to offer in framing a research agenda. The problem is not equivalent to assessing a well-defined chemical substance for which abundant data are available. An effective risk-research strategy for ENMs will require the identification of data and models to assess risks as the sparse data available are augmented. Careful planning, problem formulation, and consideration of options for managing the risks, including application of green-chemistry principles (see Box 2-1), can improve the utility of assessment for decision-making (NRC 2009).

In Table 2-1, the committee applies the framework of NRC (2009) to potential risks of ENMs. The general considerations of NRC (2009) are translated into specific considerations related to ENMs.

Challenges of Defining Potential Risks

The diverse properties of nanomaterials present a challenge to addressing potential EHS risks of ENMs. First, it is difficult to specify the composition of ENMs, because of the variety of material types and variation within types.

BOX 2-1 Incorporating Green-Chemistry Principles Into Nanomaterial Development and Application

An evolving risk-assessment process provides the best available information needed to inform regulatory decision-making and future research while providing a basis for precautionary actions that might otherwise be ruled out because of data limitations. The limitations include

- Lack of data and adequate models (for example, structure-activity and other predictive models) for nanomaterials, which results in major uncertainties in describing and quantifying nanomaterial hazard and exposure potential.
- Lack of understanding and of ability to track and keep abreast of the rapid change, already evident and expected to increase, in the array of nanomaterials and their applications.
- The diversity of nanomaterial types and variants and the poor ability to group materials for assessment purposes on the basis of known risk characteristics that can be related to specific physical properties.
- Difficulties in distinguishing between exposures and risks associated with nanoscale and conventional forms of the same substances and between naturally occurring and incidentally produced nanoscale materials and ENMs.

Nanomaterial development, informed by an evolving risk assessment, presents the opportunity to identify and reduce, at the design stage, the *inherent* potential for exposure to and the hazards of nanomaterials. Application of green-chemistry principles and design practices to nanomaterial development can help to ensure that nanomaterials are designed to minimize risk whatever their application.

ENMs seem ideally suited to such approaches, given the ability to exert precise control over composition and structure. Such atomic-scale manipulation is the defining essence of nanotechnology and is what makes it possible to impart such materials with specific properties related to function and performance. In principle, the same ability should extend to identifying and exerting control over the factors determining a nanomaterial's potential for exposure, such as persistence, mobility, or bioavailability. Similarly, it may be possible to reduce risk by reducing the inherent hazard of a nanomaterial by altering such factors as composition and reactivity. The potential to precisely define and control nanomaterial composition and structure are directly relevant to a number of Green Chemistry principles (ACS 2011) such as those addressing atom economy; use of less hazardous substances in processes; and designing for reduced toxicity, increased energy efficiency, enhanced degradation, and inherent safety.

An evolving risk-assessment process enables the identification and development of predictive tools and methods for screening nanomaterials at early stages in the development process for inherent properties that are associated with high exposure or potentially damaging biologic activity.

TABLE 2-1 Risk-Related Concerns from NRC (2009) as Applied to Nanomaterials

Topic from NRC (2009)	Consideration for nanomaterials
Emphasis should be placed on "planning and scoping" and "problem formulation" in the early phases of risk assessment to ensure that the right questions are being asked of the assessment.	Emphasis on "planning and scoping" and "problem formulation" will allow scientists and research managers to triage a wide array of materials to focus on the ones that present the greatest probability of a risk to health or the environment. For example, understanding the hierarchy of information needs from physical characteristics to potential for release to fate in the environment should allow critical early decisions in the assessment process. There may be a minimum set of information needed to address these determinants of hazard or risk for all nanomaterials, but in the near term the committee's research agenda might best focus on accumulating information on materials that appear to be reactive, likely to be released, likely to interact with other toxic materials and serve as delivery mechanisms, and likely to persist under typical environmental conditions. This somewhat simplistic example shows the importance of developing some early decision rules for implementation of the EHS research agenda. As is the nature of risk assessments, these early rules would probably be refined as experience in assessing ENMs accrues.
Refined approaches to addressing uncertainty and variability in all phases of the risk assessment from characterizing potential release through potential exposure to hazard and risk will be a critical component of information needs in this risk-research strategy.	In designing the research strategy for ENMs, a premium should be placed on a "value of information" analysis that underscores how the information gleaned from the research will be used to reduce uncertainty or to refine an appreciation of variability in exposure or risk. Methods for doing that are available and are continuing to evolve (NRC 2009).
Providing a perspective on the role of "default" values and scenarios in risk assessments will be critical.	For nanomaterials, research is needed to determine whether the traditional bases of default assumptions (for example, high to low dose, animal to human, and individual human variability in response) will apply; these issues will need to be addressed in considering approaches that lead to predicting releases or exposures that are unlikely to result in deleterious effects to humans and the environment.
Cumulative (multiple agents, same route) and aggregate (single agents, different routes) exposures need to be addressed.	For nanomaterials, research on releases of nanomaterials from multiple processes for different applications must be conducted to account for the potential for total release to the environment. Individual assessments of process-release scenarios have the potential to underestimate environmental and human exposures. Potential interactions of different nanomaterials in common disease processes should also be considered.

Countless assemblages of atoms and structures and a plethora of inorganic and organic macromolecular coatings affect their surface chemistry and therefore their behavior in the environment and their potential for biologic impact.

Second, nanoscale structures include both materials (for example, particles, fibers, or sheets) and macromolecules (for example, proteins or DNA). Many nanomaterials are particles or designed structures, not molecules. The heterogeneity of the materials profoundly affects efforts to detect or to measure the ENMs or to assess their potential to cause harm. Large biomolecules that are labeled as ENMs may be detected with high specificity using molecular recognition elements. Spectroscopic approaches may provide certifiable identification for some large molecular ENMs. Such approaches will frequently fail with the more complex structures. These materials may have highly uniform properties, while many of the more complex structures will lead to a range of possible interactions. However, the magnitude of forces and the resulting bond strengths induced by interactions with ENMs may be different from those for molecules. In addition to forces that show size dependence (for example, van der Waals interactions), the presence of a separate phase introduces surface energies and boundary effects (for example, discontinuity of crystal lattices at a particle surface and resultant surface charge) that are not present with molecules in solution.

Also, the relative impacts of kinetic compared with thermodynamic factors in controlling the environmental behavior of nanoparticles may be expected to differ from conventional chemical species for which there has been success in predicting phenomena, such as bioaccumulation or transport from, for example, use of structure-function relationships to calculate fugacity.

Third, like many "conventional" contaminants, chemical transformations of the nanomaterials and their coatings will occur in the environment and in organisms, and such transformations are not well characterized or readily predictable.

Fourth, the surface properties of nanomaterials are defined in part by the media in which they are dispersed; for example, surface water, lung fluid, salt water, and air may affect these properties differently. Because the behavior of nanomaterials may be controlled largely by surface properties, general predictions about environmental behavior and effects cannot be readily made. Overall, the lack of a clear and stable material identity makes it difficult to group materials or classes of materials that may behave similarly with respect to fate, transport, toxicity, and risk. Moreover, because most nanomaterials can be thought of not only as chemical entities but as having separate phases, there is considerable doubt regarding the appropriateness of applying or interpreting some of the conventional parameters used in exposure assessment, such as octanol-water partition coefficients and volatility.

A CONCEPTUAL FRAMEWORK LINKED TO RISK ASSESSMENT

The committee developed Figure 2-1, which establishes a conceptual framework for informing its research agenda in Chapter 5. The figure, which is not intended to portray a linear, sequential process, begins with a value-chain and lifecycle perspective. It depicts sources of nanomaterials originating throughout the lifecycle and value chain, and therefore the environmental or physiologic context that these materials are embedded in, and the processes that they affect. The circle, identified as "critical elements of nanomaterial interactions," represents the physical, chemical, and biologic properties or processes that are considered to be the most critical for assessing exposure and hazards and hence risk. Those elements exist on many levels of biologic organization, including molecular, cellular, tissue, organism, population, and ecosystem. The committee asks, What are the most important elements that one would examine to determine whether a nanomaterial is harmful? and has placed these elements at the center of the proposed research framework. The critical elements in the circle are not ordered, and the dynamic interactions among them are implied. For example, factors that affect surface affinity may also affect persistence and bioaccumulation and would not be appropriately reflected in any linear sequencing of the elements. Research needs relating to such critical elements are discussed in Chapter 3. Research priorities for addressing the critical elements are summarized in Chapter 5.

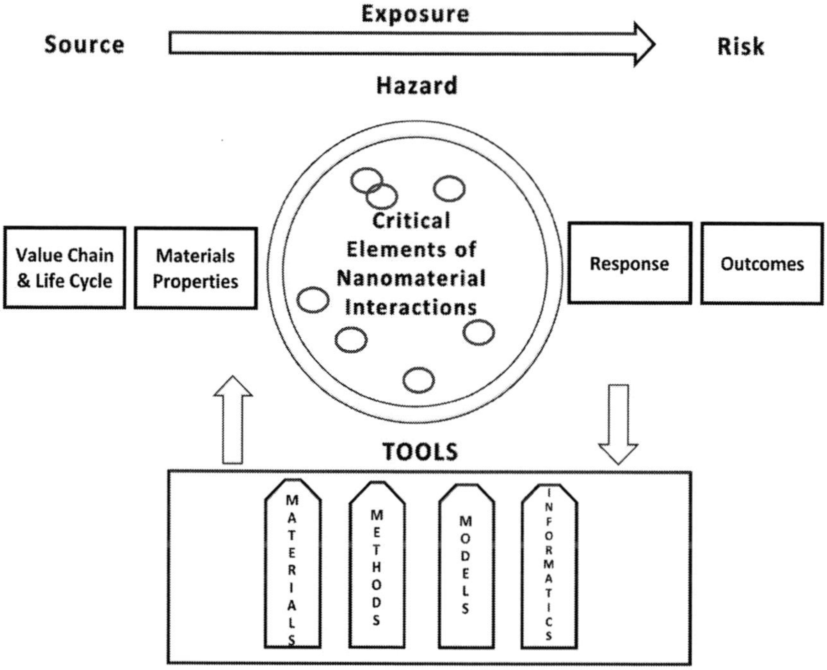

FIGURE 2-1 Conceptual framework for informing the committee's research agenda.

The lower half of the figure depicts tools needed to support an informative research agenda on critical elements of nanomaterial interactions. Improved tools will be integral products of the research agenda. The tools are materials (standardized materials that embody a variety of characteristics of interest), methods (standardized approaches for characterizing, measuring, and testing materials), models (for example, for assessing availability, concentration, exposure, and dose), and informatics (methods and systems for systematically capturing, annotating, archiving, and sharing the research results). The vertical arrows between the tools and the circle acknowledge the interplay between what is learned through research about the processes that influence exposure and hazards and the continuing evolution of the tools for carrying out research.

Inputs of nanomaterials depicted in Figure 2-1 represent releases of ENMs along the entire value chain and life cycle. Activities along the value chain imply inputs of energy and materials at each stage and the creation of waste streams. Each nanomaterial or product containing nanomaterials along the steps of the value chain has an associated life cycle of production, distribution, use, and end-of-life releases that may affect human health and the environment. The principle of including a value-chain and life-cycle perspective in the committee's conceptual framework is fundamental for assessing the risks posed by nanomaterials and is discussed in greater detail below. Understanding release mechanisms in manufacturing, transport, and product use (for example, abrasion) is implicit in this value-chain and life-cycle perspective.

A LIFE-CYCLE AND VALUE-CHAIN PERSPECTIVE WITHIN THE CONCEPTUAL FRAMEWORK

In developing the conceptual framework, the committee recognized the importance of considering aspects of the life cycle of ENMs throughout the value chain to understand the potential for exposure of humans and ecologic receptors. (See Figure 2-2, an input into the conceptual framework, Figure 2-1). The value chain extends beyond production of nanomaterials into primary and secondary products based on the parent nanomaterials. Releases can come from byproducts and wastes in addition to intended and unintended releases of the parent nanomaterials that extend throughout each step of the value chain of products that contain these materials and their life cycles. Examples of potential releases include

- Fugitive emissions of parent material.
- Process releases of nanomaterials during production and finishing of a product (for example, sawing or sanding).
- Releases during transportation or accidents.
- Releases during product or material use, recycling, recovery, or disposal.

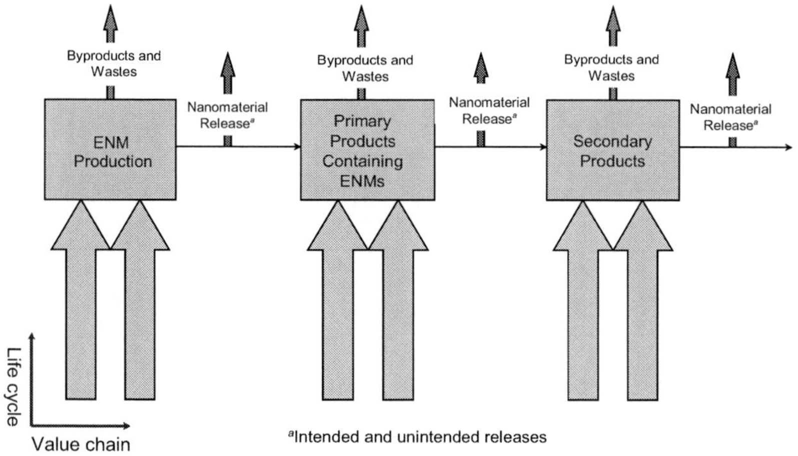

FIGURE 2-2 Potential human and ecosystem exposure through the value chain and life cycle of nanomaterial production, use, and disposal.

How nanomaterials are produced, used, reused, and disposed of largely determines the risks that they may present to human health and the environment. The risks are in two categories: risks stemming directly from exposure to nanomaterials and nanomaterial-containing products and risks produced by the "collateral damage" associated with energy consumption, material use, and wastes generated as nanomaterials are made, transported, processed, and treated for disposal.

Risks Stemming Directly from Potential Exposure to Nanomaterials

The first category of risks is derived from the potential for exposure to nanomaterials at any stage of fabrication, transport, processing, use, and end of life—activities that make up what is referred to as the life cycle of nanomaterials. The nanomaterial value chain (represented along the horizontal axis in Figure 2-2) involves the production of basic building blocks of nanomaterials and their incorporation (in later stages) into products of increasing complexity (Wiesner and Bottero 2011). For example, such ENMs as quantum dots (QDs) and single-walled carbon nanotubes (SWCNTs) might be combined as QD-SWCNT composites in primary products, such as thin films. Thin films might then be incorporated into solar cells (secondary products), which are then used in housing materials (tertiary products). Each of those products has its own life cycle associated with its fabrication, transport, processing, use, and end of life. Table 2-2 illustrates potential releases of and exposures to carbon nanotubes across the value chain and life cycle of a textile application.

TABLE 2-2 Illustration of Potential Releases of and Exposures to Carbon Nanotubes (CNTs) across the Value Chain and Lifecycle of a Textile Application

	Raw material	Manufacture 1: Materials Manufacture	Manufacture 2: Product Fabrication	Manufacture 3: Filling/ Packaging	Distribution	Use	Recycle	Disposal
CNT	Production of CNT polymers and master batches. Potential for exposure during synthesis, which may differ for each synthesis method.	Textile manufacture (next row).	Textile fabrication (next row).	Preparing CNTs for shipment to textile manufacturer. Potential for exposure during filling/packing and unpacking.	Transport of CNTs to manufacturer. Potential for release during transfer or from spills.	Use in textiles (next row); also includes epoxy resin, batteries, adhesives, and coatings.	"Recycling" of CNT raw materials may entail release during collection and re-use of remaining materials in subsequent manufacturing.	Potential for release and exposure during transport and waste management (for example, landfills, incinerators).
Product 1 (Integrating CNT into Textile)	Potential for exposure during incorporation depending on physical form and handling.	Potential for exposure during processing to make and apply a uniform material; depends on degree of automation and whether CNTs are dry, in suspension, or in masterbatch; coating of textile with CNTs could lead to	Activities include melting, spinning, weaving, sizing, knitting; bleaching, dyeing, printing, washing, drying/fixing, cutting, sewing, shaping, washing; fibre production; finishing (inspection, cleaning,	Sending CNT-treated textile to garment manufacturer.	Transport of secondary product (the garment) with CNT already incorporated into the fabric.	Use in garments (next row).	Recycling of fabric: shredding/cutting and screening, cleaning to reuse materials in new blends; release is possible from intensive treatments (for example, heat, pressure, chemical) and exposure may result from break-down or from	Disposal of unused or waste CNTs, textile scraps.

		release or exposure.	washing and packing); fibers carrying CNTs may be shed during these processes.				incorporation of CNTs into a new fabric (cross-contamination).	
Product 2 (Article of Clothing)	[N/A: CNT row].	[N/A: Product 1 row - primary product].	Pressure, chemicals, and heat of tailoring and finishing the textile may lead to release of CNTs and resulting exposure due to abrasion of fibers.	Filling/packing of secondary product (the garment).	Transport of secondary product (the garment).	Degradation of product during normal wear and tear of garment or from UV, chemicals, water, oxidation (for example, washing, ironing, heat, sweat); direct dermal exposure possible; form of released material a question: single, agglomerated ENPs or nano- or micro- scale textile containing ENP.	Textiles sent to second-hand stores or developing countries; release and exposure through wear and tear described above; recycling of fabric (previous row).	Landfills or incinerators.

Abbreviation: ENPs, engineered nanoparticles.
Sources: Chaudhry et al. 2009; EDF/DuPont 2007; Som et al. 2009.

Because of the potential for nanomaterial releases and exposures of humans or ecosystems at each stage of the value chain and life cycle, factors to consider in assessing exposure include the nanomaterial form that will be present in commercial products, the potential for the material to be released to the environment, and the transformations of the material that may affect exposure (Wiesner 2009). Analysis based on the value chain and the life cycle is rooted in an assessment of which nanomaterials are being and are expected to be produced and used.

An estimated "reservoir" of nanomaterial production can be used to obtain first-order exposure estimates that are based on explicit, easily understood assumptions regarding the quantities of nanomaterials that enter the environment integrated over the life cycle of production through disposal (Robichaud et al. 2009; Wiesner 2009; Wiesner and Bottero 2011). Understanding the fate and transport of these materials in the environment will lead to an understanding of their ability to interact with biologic systems and help in assessing risk.

Potential Risks Associated with "Collateral Damages"

The second category of risks also extends across the life cycle of nanomaterial production, use, and disposal. At each stage of the value chain (and at the links between stages of the value chain), there is consumption of energy and materials, production of wastes, and the potential for disposal, reuse, and recycling of the materials or products. Those life-cycle factors of nanomaterial production and use throughout the value chain are depicted along the vertical axis (and corresponding vertical arrows) in Figure 2-2 and may result in effects on human health and ecosystems that are independent of the nanomaterials themselves and yet are directly connected to the production of nanomaterials and the products that contain nanomaterials.

For example, the entropic penalties associated with creating order on the atomic scale indicate that energy-intensive processes will commonly be needed to produce nanomaterials (Wiesner 2009). The environmental effects of upstream energy production and use may include hazards to workers in mines, air pollution, global warming, and so on. Material use may introduce risks associated with solvent handling and disposal (Robichaud et al. 2005). It has been shown that the production of non-nanomaterial wastes from the production of carbon nanotubes (Plata et al. 2008) may pose substantial hazards. Those "collateral" risks to human health and the environment are as integral to an assessment of risks associated with nanomaterials as is the potential for exposure to and toxicity of the nanomaterials themselves. However, these factors have been largely unexamined.

An assessment of the repercussions of activities and products throughout the life cycle of production, use, disposal, and reuse of nanomaterials is needed for sustainability planning and decision-making. For any given industrial product, the life-cycle stages of resource extraction, raw-material production, product manufacturing, transportation, use, and end of life can all be associated with substantial costs and benefits to manufacturers, customers, and the environment. (See Box 2-2 for a discussion of life-cycle assessment, life-cycle inventory, and data needs.)

Although the committee recognizes that indirect collateral effects associated with the life cycle of materials and energy use in nanomaterial production may in some cases be the dominant effects on human health and the environment, the committee's research framework is focused on identifying EHS issues resulting directly from contact with nanomaterials released along the value chain and life cycle. Notably absent from the proposed framework is a consideration of important issues relating to nanomaterial fabrication, complex nanostructures and devices, and comprehensive life-cycle considerations concerning energy and materials use, reflecting a deliberate focus of this committee on nanomaterials rather than nanotechnology and a heavy emphasis on toxicologic research. However, the framework and strategy proposed by this committee address several key points raised in the NNI Signature Initiative of Sustainable Nanomanufacturing (NSET 2010b). In particular, the focus in this report on methodologic tools supports the call for novel measurement techniques. Like the NNI Initiative, the conceptual approach proposed here and the focus on nanomaterial transformations occurring after release along the value chain aligns with the NNI call for "Development of methodologies that enable accurate measurement of nanomaterial evolution and transport during product manufacturing and use, and across the material lifecycle (NSET 2010b, p. 4)."

PRINCIPLES FOR IDENTIFYING AND SETTING PRIORITIES FOR RESEARCH NEEDS IN THE CONTEXT OF THE CONCEPTUAL FRAMEWORK

One premise of the committee's framework for research is that EHS research priorities can be established on the basis of judgments regarding the relationships between nanomaterial properties and the *processes* that govern their interactions with organisms and ecosystems. The nature of the interactions will ultimately define the risk posed by the materials. The following section outlines principles that the committee considered for setting research priorities for the potential human health and environmental risks of ENMs. In many of the committee's discussions, these principles were applied implicitly as the critical research needs were considered.

Principles for Setting Priorities for Nanomaterial-EHS Research

In the paper "Towards a Definition of Inorganic Nanoparticles from an Environmental, Health and Safety Perspective," Auffan et al. (2009) illustrate how principles can be used to identify materials that are of interest from a risk

BOX 2-2 Life-Cycle Assessment, Life-Cycle Inventory, and Data Needs

Life-cycle assessment (LCA) provides a formal framework for identifying and evaluating the life-cycle effects of a product, process, or activity. Typically, effects on human health and ecosystem health and effects of pollutant deposition in all environmental media are evaluated for each stage of the life cycle, and an LCA may be performed on products at each stage of the value chain. There are many variations in LCA methods, but arguably the most broadly accepted is one formalized in the ISO-14040 series of standards (ISO 1997; Guinee et al. 2010). Often referred to as formal LCA or full LCA, the ISO method guides the quantitative assessment of environmental effects throughout a product's life cycle.

A major challenge in conducting formal LCA is to obtain reliable and available data for a life-cycle inventory (LCI). The challenge is amplified for the evolving nanomaterial industry in which production methods, markets, and patterns of product use may be unknown and confidential. Efforts have also been made to integrate consideration of social effects into LCA. The ecoefficiency assessment of BASF corporation has recently been extended to include social effects (Schmidt et al. 2005). Individual indicators of a product's effects on human health and safety, nutrition, living conditions, education, workplace conditions, and other social factors are assessed and scored relative to a reference (usually the product being replaced).

Although LCA based on a robust LCI may prove to be a useful tool in assessing EHS risks posed by manufactured nanomaterials, it must be remembered that releases to the environment, representing an upper bound on *potential* for exposure, will not equate to actual exposure of humans or ecologic receptors. Fate, transport, and transformation processes of nanomaterials in the environment need to be considered. Data needs include

- Characterizing commonly used nanomaterials.
- Understanding the potential for release of nanomaterials throughout the life cycle of the material and the value chain leading to products.
- Placing potential releases into an exposure context.
- Providing bases for assessing risk to human health and the environment.

In addition, a broader framework that combines life-cycle assessment and risk analysis may help to inform our understanding of potential risks and environmental impacts of ENMs (Evans et al. 2002; Matthews et al. 2002; Shatkin 2008).

Current knowledge needs to be assessed and a gap analysis performed to understand critical research and data needs for addressing the EHS aspects of nanomaterials (see Chapter 3). Addressing the issues of modeling vs monitoring—for example, releases, fate and transport, exposure, dose, and potential effects—will be critical for the success of this effort (see discussion in Chapter 4).

perspective. Regarding important risk-related characteristics of ENMs, Auffan et al. considered developing a risk-based definition of inorganic nanoparticles that is founded on novel size-dependent properties. Contrary to the title of their paper, Auffan et al. pose a set of principles for identifying materials of interest rather than a rigid definition for classifying ENMs. The science-based approach that they adopted allows materials presenting new or unusual risks to be distinguished from materials that present more conventional risks. Their approach establishes criteria for determining the probability that a material measuring 1-100 nm will exhibit novel properties that might lead to new or unusual risks.

Building on that idea, the present committee focuses on a set of principles in lieu of definitions to help identify nanomaterials and associated processes on which research is needed to ensure the responsible development and use of the materials. The principles were adopted in part because of concern about the use of rigid definitions of ENMs that drive EHS research and risk-based decisions (Maynard 2011; Maynard et al. 2011a). The principles are technology-independent and can therefore be used as a long-term driver of nanomaterial risk research. They help in identifying materials that require closer scrutiny regarding risk irrespective of whether they are established, emerging, or experimental ENMs. The principles are built on three concepts: emergent risk, plausibility, and severity; the principles are based on proposals articulated by Maynard et al. (2011b).

Emergent risk, as described here, refers to the likelihood that a new material will cause harm in ways that are not apparent, assessable, or manageable with current risk-assessment and risk-management approaches. Examples of emergent risk include the ability of some nanoscale particles to penetrate to biologically relevant areas that are inaccessible to larger particles, the failure of some established toxicity assays to indicate accurately the hazard posed by some nanomaterials, scalable behavior that is not captured by conventional hazard assessments (such as behavior that scales with surface area, not mass), and the possibility of abrupt changes in the nature of material-biologic interactions associated with specific length scales. Identifying emergent risk depends on new research that assesses a novel material's behavior and potential to cause harm.

Emergent risk is defined in terms of the potential of a material to cause harm in unanticipated or poorly understood ways rather than being based solely on its physical structure or physicochemical properties. Thus, it is not bound by rigid definitions of nanotechnology or nanomaterials. Instead, the principle of emergence enables ENMs that present unanticipated risks to human health and the environment to be distinguished from materials that probably do not. It also removes considerable confusion over how nanoscale atoms, molecules, and internal material structures should be considered from a risk perspective, by focusing on behavior rather than size.

Many of the ENMs of concern in recent years have shown a potential to lead to emergent risks and would be tagged under this principle and thus require further investigation. But the concept also allows more complex nanomaterials to be considered—those in the early stages of development or yet to be developed. These include active and self-assembling nanomaterials. The principle does raise the question of how "emergence" is identified, being by definition something that did not exist previously. However the committee recognized that in many cases it is possible to combine and to interpret existing data in ways that indicate the possible emergence of new risks. For example, some research has suggested that surface area is an important factor that affects the toxic potency of some ENMs; ENMs that have high specific surface area and are poorly soluble might pose an emergent risk.

Plausibility refers in qualitative terms to the science-based likelihood that a new material, product, or process will present a risk to humans or the environment. It combines the possible hazard associated with a material and the potential for exposure or release to occur. Plausibility also refers to the likelihood that a particular technology will be developed and commercialized and thus lead to emergent risks. For example, the self-replicating nanobots envisaged by some writers in the field of nanotechnology might legitimately be considered an emergent risk; if it occurs, the risk would lie outside the bounds of conventional risk assessment. But this scenario is not plausible, clearly lying more appropriately in the realm of science fiction than in science. The principle of plausibility can act as a crude but important filter to distinguish between speculative risks and credible risks.

The principle of *severity* refers to the extent and magnitude of harm that might result from a poorly managed nanomaterial. It also helps to capture the reduction in harm that may result from research on the identification, assessment, and management of emergent risk. The principle offers a qualitative reality check that helps to guard against extensive research efforts that are unlikely to have a substantial effect on human health or environmental protection. It also helps to ensure that research that has the potential to make an important difference is identified and supported.

Together, those three broad principles provide a basis for developing an informed strategy for selecting materials that have the greatest potential to present risks. They can be used to separate new materials that raise safety concerns from materials that, although they may be novel from an application perspective, do not present undetected, unexpected, or enhanced risks. They contribute to providing a framework for guiding a prioritized risk-research agenda. In this respect, the principles were used by the committee as it considered the pressing risk challenges presented by ENMs.

When the principles are applied to existing and emerging ENMs, various groups of materials that may warrant further study are evident. Those groups, identified below, are not intended to be comprehensive, but they are the basis for beginning to map out material properties that need to be addressed in a risk-research strategy (Maynard et al. 2011b).

- *Materials that demonstrate abrupt scale-specific changes in biologic or environmental behavior.* Materials that undergo rapid size-dependent changes in physical and chemical properties that affect their biologic or environmental behavior may pose a hazard that is not predictable based on what is known about larger-scale materials of the same composition.

- *Materials capable of penetrating to normally inaccessible places.* Materials that, on the basis of their size or surface chemistry or both, are able to persist in or penetrate to places in the environment or body that are not accessible to larger particles of the same chemistry may present emergent risks. If there is a credible scenario for accumulation of, exposure to, or an organ-specific dose of a nanomaterial that is not expected according to the behavior of the dissolved material or larger particles of the same material, a plausible and emergent risk is possible.
- *Active materials.* Materials that change their biologic behavior in response to their local environment or a signal present dynamic risks that are not well understood. Active materials might include materials whose surface charge leads to association with other materials in the environment, which allows the nanomaterial to function as an efficient delivery system for potentially toxic materials, such as metals and polyaromatic hydrocarbons. Active materials might also include materials whose enzymatic or catalytic processes pose a potential hazard in biologic systems. In addition, it is plausible that nanomaterials that have a three-dimensional structure, similar to natural ligands, could activate receptor-mediated processes in humans and the environment.
- *Self-assembling materials.* Materials that are designed to assemble into new structures in the body or the environment on release pose issues that may not be captured by current risk-assessment approaches.
- *Materials exhibiting a scalable hazard that is not captured by conventional dose metrics.* When hazard scales according to parameters that are not typically used in risk assessment, emergent risks may arise because dose-response relationships may be inappropriately quantified. For example, the hazard presented by an inhaled material may scale with the surface area of the material, but if risk assessment is based on mass, the true hazard may not be identified; the material has the possibility of causing unexpected harm.

Applying the Principles to the Value Chain and Life Cycle of Nanomaterials and Products

The principles can be applied to both the value chain of materials and products and their life cycle to identify context-specific risks that may arise and require further research to assess and manage them. The concepts of plausibility, emergence, and severity can help to differentiate between what may be considered more and less important risks. For example, generating and handling multiwalled carbon nanotubes in a workplace—materials that have demonstrated novel properties that include, for example, strength and electric conductivity—may present a plausible and emergent risk. It is only recently that production of these materials has started commercially; there are indications that some forms of carbon nanotubes are more harmful than their carbon base might indicate; and there is a potential for human exposure (Maynard et al. 2004; Han et al. 2008; Evans et al. 2010). However, riding a bicycle that incorporates multiwalled nanotubes in the frame or using a cellular telephone with a battery containing small quantities of nanotubes is unlikely to lead to important exposure. In those cases, although the emergent risk might remain, the plausible risk is much reduced; nevertheless, when the products are disposed of or prepared for recycling, a plausible and possibly severe risk may re-emerge as the material again becomes potentially dispersible and biologically available.

Those examples demonstrate how the principles of plausibility, emergent risk, and severity allow important risks or "hot spots" to be identified over the value chain and life cycle of the material. The principles provide a systematic basis for identifying and setting priorities among properties of nanomaterials as research subjects in addressing risks.[1]

Criteria for Selecting Research Priorities

Each of the above types of materials (they are not exclusive), illustrates key research questions that need to be addressed if emergent and plausible risks are to be identified, characterized, assessed, and managed. The principles described above can be applied to set priorities for the study of ENMs. However, a comprehensive research strategy also will address both near-term and long-term issues regarding the EHS aspects of nanomaterials, including identifying the properties of ENMs that make them potentially hazardous; determining how to harmonize collection and storage of pertinent but diverse data types to enable risk-assessment modeling and risk management; developing new tools to measure ENMs in complex environmental and biologic matrices and to model exposure and hazard pathways; and identifying justifiable simplifications that can reduce the level of complexity to enable comprehensive risk assessment of ENMs. And it should outline a path to address complex mixtures of ENMs, to understand their transformations and interactions with existing environmental contaminants, and to assess how the transformations and interactions affect their behavior and effects.

In addition to the issues of life-cycle and value-chain perspective discussed earlier, the committee identified the following criteria as a basis of setting priorities for research:

- Research that advances knowledge of both exposure and hazard wherever possible.

[1]A similar definition-independent approach to addressing potential risks arising from ENMs has previously been proposed in the Nano Risk Framework developed by the Environmental Defense Fund and DuPont (Environmental Defense/DuPont 2007).

• Research that leads to the production of risk information needed to inform decision-making on nanomaterials in the market place.

• Research efforts to address short-term needs that serve as a foundation for moving beyond case-by-case evaluations of nanomaterials and allows longer-term forecasting of risks posed by newer materials expected to enter commerce.

• Research that promotes the development of critical supporting tools, such as measurement methods, limitations of which hinder the conduct of research in processes that control hazards and exposure.

• Research on ecosystem-level effects that addresses exposure or hazard scenarios that are underrepresented in the current portfolio of nanotechnology-related EHS research; for example, impacts on ecosystem processes and on organisms representing different phyla and environments.

REFERENCES

ACS (American Chemical Society). 2011. The Twelve Principles of Green Chemistry. ACS Green Chemistry Institute [online]. Available: http://portal.acs.org/portal/acs/corg/content?_nfpb=true&_pageLabel=PP_ARTICLEMAIN&node_id=1415&content_id=WPCP_007504&use_sec=true&sec_url_var=region1&__uuid=b0b83343-c387-486e-8d4a-d30de190775e [accessed Oct. 25, 2011].

Auffan, M., J. Rose, J.Y. Bottero, G.V. Lowry, J.P. Jolivet, and M.R. Wiesner. 2009. Towards a definition of inorganic nanoparticles from an environmental, health and safety perspective. Nat. Nanotechnol. 4(10):634-641.

Chaudhry, Q., R. Aitken, S. Hankin, K. Donaldson, S. Olsen, A. Boxall, I. Kinloch, and S. Friedrichs. 2009. Nanolifecycle: A Lifecycle Assessment Study of the Route and Extent of Human Exposure via Inhalation for Commercially Available Products and Applications Containing Carbon Nanotubes. Final Report. The Food and Environmental Research Agency (FERA), York, UK [online]. Available: http://www.man.dtu.dk/English/About/personer.aspx?lg=showcommon&id=265562 [accessed Mar. 18, 2011].

Dreher, K.L. 2004. Health and environmental impact of nanotechnology: Toxicological assessment of manufactured nanoparticles. Toxicol. Sci. 77(1):3-5.

Drezek, R.A., and J.M. Tour. 2010. Is nanotechnology too broad to practise? Nat. Nanotechnol. 5(3):168-169.

EDF/DuPont (Environmental Defense Fund and DuPont). 2007. Nano Risk Framework. Environmental Defense Fund, Washington, DC, and DuPont, Wilmington, DE. June 2007 [online]. Available: http://nanoriskframework.com/page.cfm?tagID=1081 [accessed Mar. 17, 2010].

Evans, J.S., P. Hofstetter, T.E. McKone, J.K. Hammitt, and R. Lofstedt. 2002. Introduction to special issue on life cycle assessment and risk analysis. Risk Anal. 22(5):819-820.

Evans, D.E., B.K. Ku, M.E. Birch, and K.H. Dunn. 2010. Aerosol monitoring during carbon nanofiber production: Mobile direct-reading sampling. Ann. Occup. Hyg. 54(5):514-531.

Guinee, J.B., R. Heijungs, G. Huppes, A. Zamagni, P. Masoni, R. Buonamici, T. Ekvall, and T. Rydberg. 2010. Life cycle assessment: Past, present, future. Environ. Sci. Technol. 45(1):90-96.

Han, J.H., E.J. Lee, J.H. Lee, K.P. So, Y.H. Lee, G.N. Bae, S.B. Lee, J.H. Ji, M.H. Cho, and I.J. Yu. 2008. Monitoring multiwalled carbon nanotube exposure in carbon nanotube research facility. Inhal. Toxicol. 20(8):741-749.

ISO (International Organization for Standardization). 1997. Environmental Management-Life Assessment-Principles and Framework. ISO 14040. International Organization for Standardization, Switzerland.

Matthews, H.S., L. Lave, and H. MacLean. 2002. Life cycle impact assessment: A challenge for risk analysis. Risk Anal. 22(5):853-859.

Maynard, A.D. 2011. Don't define nanomaterials. Nature. 475(7354):31.

Maynard, A.D., P.A. Baron, M. Foley, A.A. Shvedova, E.R. Kisin and V. Castranova. 2004. Exposure to carbon nanotube material: Aerosol release during the handling of unrefined single-walled carbon nanotube material. J. Toxicol. Environ. Health 67(1):87-107.

Maynard, A.D., D. Bowman, and G. Hodge. 2011a. The problem of regulating sophisticated materials. Nat. Mater. 10(8):554-557.

Maynard, A.D., D.B. Warheit, and M.A. Philbert. 2011b. The new toxicology of sophisticated materials: Nanotoxicology and beyond. Toxicol. Sci. 120(Suppl. 1):S109-S129.

NRC (National Research Council). 1983. Risk Assessment in the Federal Government: Managing the Process. Washington, DC: National Academy Press.

NRC (National Research Council). 2009. Science and Decisions: Advancing Risk Assessment. Washington, DC: National Academies Press.

NSET (Nanoscale Science, Engineering, and Technology Subcommittee). 2010a. FAQs: Nanotechnology. National Nanotechnology Initiative. Nanoscale Science, Engineering, and Technology Subcommittee [online]. Available: http://www.nano.gov/nanotech-101/nanotechnology-facts [accessed July 13, 2011].

NSET (Nanoscale Science, Engineering, and Technology Subcommittee). 2010b. Sustainable Nanomanufacturing-Creating the Industries of the Future, Final Draft, July 2010. NSTC Committee on Technology. Subcommittee on Nanoscale Science, Engineering, and Technology [online]. Available: http://www.nano.gov/sites/default/files/pub_resource/nni_siginit_sustainable_mfr_revised_nov_2011.pdf [accessed Nov. 3, 2011].

Plata, D.L., P.M. Gschwend, and C.M. Reddy. 2008. Industrially synthesized single-walled carbon nanotubes: Compositional data for users, environmental risk assessments, and source apportionment. Nanotechnology. 19(18):185706.

Robichaud, C.O., D. Tanzil, U. Weilenmann, and M.R. Wiesner. 2005. Relative risk analysis of several manufactured nanomaterials: An insurance industry context. Environ. Sci. Technol. 39(22):8985-8994.

Robichaud, C.O., A.E. Uyar, M.R. Darby, L.G. Zucker, and M.R. Wiesner. 2009. Estimates of upper bounds and trends in nano-TiO_2 production as a basis for exposure assessment. Environ. Sci. Technol. 43(12):4227-4233.

Schmidt, I., M. Meurer, P. Saling, A. Kicherer, W. Reuter, and C.O. Gensch. 2005. SEEbalance®: Managing sustainability products and processes with the Socio-Eco-Efficiency Analysis by BASF. Greener Management International. 45(July):79-94.

Shatkin, J.A. 2008. Informing environmental decision making by combining life cycle assessment and risk analysis. J. Ind. Ecol. 12(3):278-281.

Som, C., M. Halbeisen, and A. Köhler. 2009. Integration of Nanoparticles in Textiles [in German]. Textile Federation of Switzerland and EMPA, Switzerland. January. 5, 2009 [online]. Available: http://www.empa.ch/plugin/template/empa/*/78231 [accessed Mar. 18, 2011].

Wiesner, M.R. 2009. Life Cycle of Nanoproducts. Nanostructured Materials and Non-manufacturing Definitions: Data Gaps and Research Needs. Presentation at Nanotechnology and Life Cycle Analysis Workshop, November 5-6, 2009, Chicago, IL [online]. Available: http://www.uic.edu/orgs/nanolcaworkshop/wiesner.html [accessed July 26, 2011].

Wiesner, M.R. and J.Y. Bottero. 2011. A risk forecasting process for nanostructured materials, and nanomanufacturing. Comptes Rendus Physique. 12(7):659-669.

3

Critical Questions for Understanding Human and Environmental Effects of Engineered Nanomaterials

INTRODUCTION

This chapter articulates the most pressing research gaps to be addressed for advancing understanding of the environmental and human health effects of nanotechnology with the overall goal of mitigating any risk. The gaps, which are articulated as questions, are organized according to the source-response paradigm that runs through this report and are evaluated by the principles established in Chapter 2. These questions relate to engineered nanomaterial (ENM) sources and manufacturing; modifications, fate, and transport; bioavailability and dose; and effects on organisms and ecosystems. Figure 3-1 presents a source-to-response paradigm that the committee used to organize and to identify the gaps and corresponding critical research questions. The boxes above the arrow generally track the life cycle of an ENM. The topics below the arrows are specific issues that help to define the research landscape.

The source-response paradigm is familiar, but in extending it to nanotechnology several specific elements need to be included. Most notable is the challenge of identifying ENM sources (far left box). Because uses of nanomaterials are relatively new, changing, and expanding rapidly, definitive information about most exposure scenarios is not available.

Figure 3-1 also highlights the central role of modifications and fate and transport of ENMs in determining exposure (second box). As ENMs move from a source to a biologic receptor, myriad modifications can occur, which are challenging to anticipate and characterize. As a result, the measurement and definition of dose and bioavailability can be difficult, and these are listed as a separate research subject (third box).

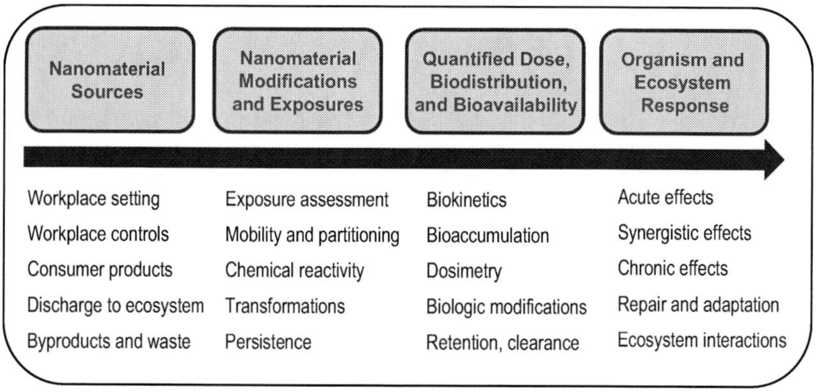

FIGURE 3-1 Central topics for EHS research on ENMs. Research on the EHS aspects of nanotechnology can be organized into groupings that map onto a framework that considers how a source of nanomaterial (left) may result in an organism or ecosystem response (right).

Although Figure 3-1 presents a paradigm for organizing information about nanotechnology and its risks, it does not address the full diversity of exposed populations. Occupational exposure to ENMs is likely, given the extensive research enterprise and burgeoning startup business community. Inhalation exposure in manufacturing may occur if processes rely on gas-phase production of materials or if materials are aerosolized. Consumer exposure to ENMs also is of immediate interest in that makers of products ranging from sunscreens to car bumpers have touted the inclusion of nanotechnology (PEN 2011). Topical and ingestion exposure from use of personal-care and other consumer products is possible. And the environment is exposed through disposal or intentional application of ENMs for remediation and through incidental or accidental release and runoff. However, those different exposure scenarios involve many common research issues, particularly in the early stages of the ENM life cycle. Because research on risks to human vs ecosystem health poses different challenges, particularly when ENM-related hazards are considered, the discussion in some sections is separated to reflect these differences.

The discussion below is organized according to the source-to-response paradigm to address critical gaps, but many key questions in nanotechnology-EHS research are intrinsically systems problems that can be addressed only by integrating the interactions of various components of the paradigm (See Figures 2-1 and 3-1). The assumptions and data from the more established foundation disciplines of pulmonary toxicology, environmental impact analysis, nanomedicine, and risk assessment are discussed where relevant. The chapter concludes with a compilation of research questions based around Figure 3-1 (see Table 3-1); the questions capture issues that are critical to the many stakeholders responsible for managing potential ENM-related risks.

PRIOR RESEARCH-GAP ANALYSIS—AN OVERVIEW

Several convergent themes can be found in past research-gap analyses of the environmental and health impacts of nanotechnology (see Chapter 1). Most notable for this analysis are

- The vital need for standardized ENMs, harmonized characterization methods, and standard biologic tests.
- Research gaps concerning the in vivo evaluation of ENMs, particularly for chronic exposures and their impact on physiologic or biochemical endpoints.[1]
- Gaps in understanding low-level environmental exposure to ENMs and their impact on organisms through changes in development, reproduction, and growth.

While this is not a complete list, those topics represent conclusions reached in multiple synthesis reports over the last few years.

The need for standardization has emerged repeatedly in research-needs discussions and reflects the communitywide sentiment that ensuring reproducible and meaningful findings requires a common platform of materials, methods, and, most recently, models. In 2002, EPA held a workshop on nanotechnology and the environment that included a discussion of impacts; that event and a workshop at the University of Florida began the analysis of the grand challenges in this field (EPA 2002). Those early efforts emphasized the need for uniform and standard materials to facilitate comparison of results between different exposure and toxicity studies. Later workshops echoed earlier findings and emphasized the need to harmonize protocols for toxicologic evaluations. Testimony given to the U.S. Congress on strategies for nanotechnology-EHS research also highlighted the need to standardize materials, methods, and models (for example, Denison 2005). Chapter 4 addresses these needs.

Some workshops have offered priorities for research. In 2006, a meeting at the Woodrow Wilson Center considered the toxicology of ENMs by route of exposure and noted the importance of dermal and gastrointestinal exposure of humans (Balbus et al. 2007). That emphasis reflected the perception that the substantial literature on the toxicology of inhaled particles in humans could inform understanding of that exposure route and that comparatively little was known about the potential for exposure to ENMs by the dermal and gastrointestinal routes. Also noted in many reports is the importance of chronic-toxicity and developmental-toxicity studies and the overemphasis on acute studies (for example, Hirose et al. 2009). This emphasis reflects the relative ease of performing acute studies in vitro, providing faster, lower-cost studies that dominate publications in the peer-reviewed literature.

---

[1]Physiologic or biochemical changes resulting from exposures.

Additional insights into critical research gaps can be derived from examination of the peer-reviewed literature. The International Council on Nanotechnology (ICON) maintains a database of peer-reviewed publications on nanotechnology and issues related to nanotechnology-EHS research. The database integrates the complementary journal content found in the Web of Knowledge and PubMed. For this assessment, ICON's categorization of the publications is critical. Skilled researchers review publication abstracts that meet a broad set of criteria and then classify the publications on the basis of their content—for example, exposure and hazards, environmental vs human health outcomes, and types of materials. An analysis of ICON's database of peer-reviewed publications reveals the relative imbalance between research on exposure and research on hazards: exposure-assessment studies constitute fewer than 25% of all papers published between 2001 and 2009 (Figure 3-2). This gap is important to address, but reflects a common trend for all chemicals in that greater attention is given to toxicity research than to exposure research. Similarly, there is a dearth of information about workplace exposure.

There is also an imbalance between environmental and human health studies (Figure 3-3).

A theme observed both in the recent NNI workshops (NNI 2011a) and in the research directions of recently funded centers (NNI 2011b) has been the importance of systematic modeling of the relationships between materials and effects. The focus on systematic modeling has developed as investigators have grappled with the challenges of managing many types of nanomaterials, of formulations, of surface coatings, of delivery or packaging systems, and of exposure routes. The sheer number of possible variations makes conventional testing paradigms impractical. The role of modeling is addressed in detail in Chapter 4.

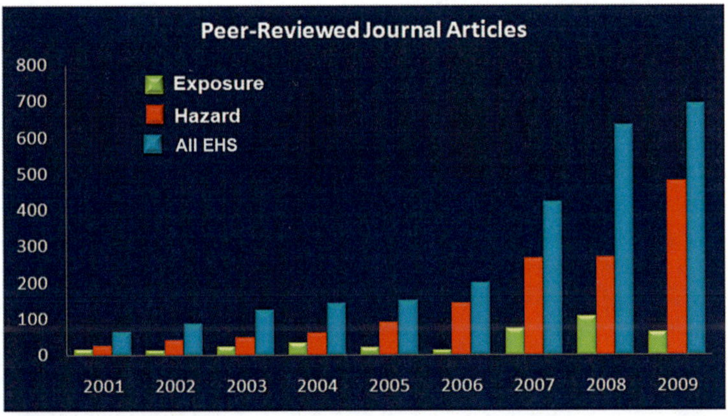

FIGURE 3-2 The number of peer-reviewed publications relating to exposure and hazard. Although the number of peer-reviewed publications on EHS effects of nanotechnology has grown substantially, far more publications address issues related to hazard than exposure. Adapted from ICON 2011.

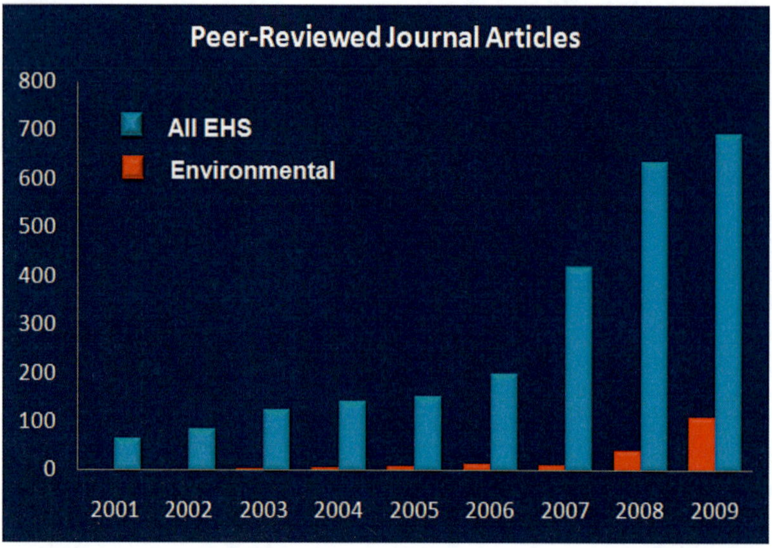

FIGURE 3-3 The number of peer-reviewed publications on environmental issues. Although the number of peer-reviewed publications on EHS effects of nanotechnology has grown substantially, only a small number of publications address environmental issues. Adapted from ICON 2011.

RESEARCH-GAP ANALYSIS AND IDENTIFICATION OF CRITICAL RESEARCH QUESTIONS

Figure 3-1 is used in this section to structure consideration of the central research questions with major issues highlighted in boldface and mostly formulated as research questions. The committee addresses the relevance of these central questions to an understanding of potential human and environmental effects of ENMs.

Sources of Engineered Nanomaterials

A major issue related to EHS consequences of nanotechnology is the uncertainty of potential exposure. Nanotechnology is not yet highly developed as an industry, so that there is little experience with actual exposures to workers, the population, and the environment generally. Moreover, given the expected growth of the industry, existing release scenarios may not be indicative of those in the future (Figure 3-4). This uncertainty about exposure scenarios complicates problem definition and scoping—the necessary first step in a risk assessment. Efforts are needed to obtain exposure data related to present conditions in order to characterize exposures, and in combination with hazard data, assess potential risks.

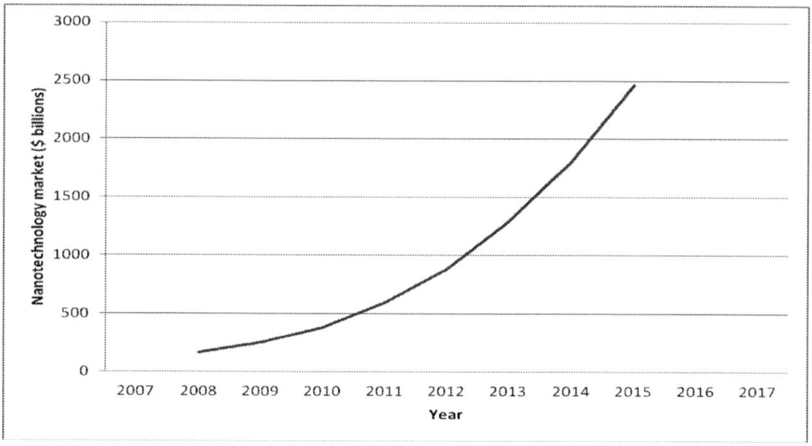

FIGURE 3-4 Projection of the size of the nanotechnology market. Source: Data from Lux 2009.

Table 3-2 lists a selection of classes of ENMs that notably include many variants. These materials represent different systems for researchers to characterize and study, and a consistent question facing researchers is, ***which of these classes of materials is of immediate interest and of relevance for EHS research?*** One approach to answer that question uses projected market size and potential risk based on plausibility and emergent risk (see Chapter 2 for discussion of these terms.) Such an analysis has led, for example, to a research focus on nanoscale silver. Nanosilver is widely used in commerce and when released into the environment could have effects on aquatic life (J.M. Johnston et al. 2010).

However, because the number of products containing nanoscale materials is expected to explode in the next several years (see Figure 3-4), selecting target materials on the basis of existing or projected market size is problematic. In addition, some products may result in very small releases (for example, computer devices) and other products greater releases to particular populations (for example, cosmetics). Research could be directed toward three or four classes of materials or materials in specific types of applications (for example, cosmetics) that pose a plausible and emergent risk. Research also could be focused on fundamental processes affecting exposure potential, for example, factors affecting releases from commonly used nanomaterial-containing matrices such as plastics, or on fundamental properties influencing nanoparticle-macromolecular interactions. The research on specific material types should be continually revisited and informed through regular surveys of nanomaterial production and use patterns.

What are the maximum anticipated amounts of exposures to ENM sources to which workers, consumers, and ecosystems could be exposed? Realistic estimates of human and environmental exposures to ENMs from well-characterized sources are critical inputs for setting priorities for research. Surveys and registries of known products and their ENM constituents could allow ENM users, industry, and academic researchers to characterize at least the maximum concentrations of ENMs of various types from sources, particularly in workplace environments. Accurate information on these ENM sources is vital, as researchers today make crude assumptions about release potential and ENM concentrations in workplaces and other environments. Characterizing the nature of point sources of ENMs (for example, wastewater-treatment plant effluent) and nonpoint sources of ENMs (for example, stormwater or agricultural runoff) is essential for determining environmental compartments and locations that will be affected and for estimating the expected concentration of ENMs in those media.

How might concentrations of ENMs from different sources apportion themselves in workplace, consumer, and various environmental compartments? Although basic information about point sources is essential, it is clear that ENMs will not remain at their sources. They will move into different environmental compartments, and environmental-exposure models are needed to describe this behavior. Methods for estimating releases of ENMs to the environment are currently based on estimates of total material flows (Blaser et al. 2008; Mueller and Nowack 2008; Robichaud et al. 2009; Gottschalk et al. 2010) and on assumptions of distribution of products that they will be used in and the fraction of the ENMs in those materials that will be released over their life cycle. Current models do not incorporate information about ENM properties. Assumptions about product uses and fractions of ENMs in materials that are released are empirical at best and not readily validated as there is presently no accurate means of tracking the mass of ENMs produced, used in specific products, or disposed of. More important, there is no information on the fraction of

TABLE 3-2 Examples of Common Nanoscale Materials and Their Applications[a]

	Features and Types	Example Products
Fullerenes	C_{60}, carbon nanotubes, graphene	Conductive films, fuel cells, composites, cosmetics
Ceramics	Iron oxides, ceria, titania	Photocatalysts, magnetic data storage, window coatings, sun-screens, paint
Metals	Silver, gold, platinum	Antimicrobial fabrics, oxidation catalysts, sensor elements
Quantum Dots	Cadmium chalcogenides	Solar cells, diodes, biologic markers
Polymers	Copolymer assemblies, dendrimers	Coatings, rheologic control, drug delivery

[a]There is little information on the relative exposure to these different materials or their products.

ENMs that may be expected to be released during normal use, at the end of a material's life, or during recycling. The models mentioned above are designed to provide an upper limit of ENM exposure, but they are representations of still largely unknown scenarios. Particularly for environmental exposure, detection schemes are needed to validate the models and to signal that there is a potential for exposure. That kind of early-warning system could also be developed for workplace exposure. ***Thus, an important research question is, How can ENMs be detected in air, in water, and in complex media, to allow real-time monitoring of sources of ENM exposure?***

Modification of and Exposure to Engineered Nanomaterials

Identification of a source of ENMs leads to the need to assess the potential for exposure. For ENMs, the assessment of potential exposures is complicated by the many modifications of the ENMs that may occur. In addition to investigating how much nanomaterial may be present at a receptor, it is also critical to specify the form of the nanomaterial at the point of exposure. A nanoscale material may undergo both subtle and extreme changes as it moves through biologic and environmental systems. The changes can be in size, surface chemistry, and reactivity and these changes might lead to different hazards. That complexity is not dissimilar to exposure issues related to, for example, dissolution of metals in water; depending on the details of the water chemistry, metals may be in different oxidation states or have different degrees of bioavailability (Mahendra et al. 2008). As in the case of metals, models that can predict the form of nanoscale materials, given the environmental compartments, are vital. Once the form of the nanomaterial is established, many of the exposure questions are reduced to accurate measurement of the quantity of material. In addressing ENM modifications and the related implications for exposure and hazard, the distinction between human health and environmental health is relevant. Modification processes may be different if a material is first transported through the environment vs through the human body, and the tools needed for exposure assessment also differ and are treated separately below.

Human Health—Needed Research on Material Modification and Exposure Assessment

There are three primary routes of human exposure: inhalation, ingestion, and dermal absorption. Inhalation is the most studied pathway; research on the effects of inhaling particles in the ultrafine size range long antedated the emergence of nanotechnology, and commercial instruments are available for detecting submicrometer ambient particles. For example, when measuring airborne engineered nanoparticles, equipment such as the Scanning Mobility Particle Sizer or Fast Mobility Particle Sizer can be used (McMurry 2000; Asbach et al. 2009; Jeong and Evans 2009; Aggarwal 2010). However, the circumstances for inhalation exposure to ENMs have not been well characterized and one critical research gap is the identification of ***conditions that will cause ENMs that are in the gas phase, in liquids, or embedded in solids to become airborne.***

The frequency of conditions that might lead to inhalation exposures over the lifecycle of ENMs is also uncertain. While ENMs manufactured in the gas phase can produce aerosolized materials, most applications use ENMs in solution or embedded in devices or composites. However, application-specific processes could result in inhalation exposures. Nanocomposites could be machined so that dust is produced, or they may be used in applications (for example, as additives to fuels) that lead to their emission in exhausts or gases. Only a few studies have assessed exposure at nanomaterial manufacturing sites (Kuhlbusch et al 2004; Maynard et al 2004; Bello et al. 2008; Han et al 2008; Bello et al. 2009; Tsai et al. 2009; Lee et al. 2010; Methner et al 2010; Sahu and Biswas 2010). The potential for such exposures at other stages of the ENM lifecycle needs more study.

When airborne particles are inhaled, they deposit in three regions of the respiratory tract (upper, tracheobronchial, and alveolar) depending on their size. For example, the smaller nanoparticles (less than 5 nm), if inhaled as single particles, deposit to a high degree in the upper respiratory tract (nasopharyngeal region) whereas larger particles (about 10 nm) deposit to a greater degree in the tracheobronchial region, and the larger particles (about 20 nm) deposit with the highest deposition efficiency (up to 50% of inhaled particles) in the alveolar region. The primary mechanism for deposition of airborne nanomaterials is by diffusion, while larger particles, including aggregated and agglomerated nanoparticles, deposit by sedimentation (gravitational forces) and by impaction (inertia).[2]

During inhalation, modifications of ENMs may occur in the lung lining fluid, possibly including agglomeration and deagglomeration. The size dependence of particles for clearance by airway mucociliary and alveolar mechanisms and translocation from the respiratory tract into the bloodstream has been examined in model systems (Guo et al. 2007; Kreyling et al. 2002, 2009; Semmler-Behnke et al. 2007, 2008; Tang et al. 2009). The agglomeration-aggregation[3] state of ENMs deep in the respiratory tract is not well understood and is related to the modifications of the ENM surface induced by the lung lining fluid and along translocation pathways.

---

[2]Other minor deposition mechanisms specific for certain materials include interception (for fibers), electrostatic image forces, and condensational growth (in the highly saturated regions of the respiratory tract).

[3]Agglomeration results from "the collection of weakly bound particles or aggregates or mixtures of the two where the resulting external surface area is similar to the sum of the surface areas of the individual components" (ISO 2008). Aggregation results from "strongly bonded or fused particles where the resulting external surface area may be significantly smaller than the sum of calculated surface areas of the individual components" (ISO 2008). However, the committee recognizes that the distinction between these terms is more operational than theoretical, and different communities use these terms differently.

Far less is known about dermal and ingestion exposure than about inhalation exposure. A number of review articles have examined the exposure to nanomaterials via the dermal route (for example, Schneider et al. 2009; Smijs and Bouwstra 2010; Prow et al. 2011). The ability of nanomaterials to penetrate skin is influenced by the condition of the skin and the physicochemical properties of the nanomaterials (for example, size, charge density, photostability, and hydrophobicity). Data suggest that nanoparticles greater than 10 nm in diameter are unlikely to penetrate human skin. However, uptake may occur if skin is damaged or diseased (Mortensen et al. 2008; Prow et al. 2011), although data on penetration of nanomaterials into damaged skin is limited. ***The persistence*[4] *and potential effects of ENMs, particularly photoactive materials, on skin requires additional research.***

Research on ingestion of ENMs as a direct exposure route is just beginning. The Food and Agriculture Organization and the World Health Organization (FAO/WHO 2009) summarized information on the potential food safety implications of ENMs. Direct exposure to ENMs may occur from their use in food, for example to enhance nutritional value or to improve flavor or color (EFSA 2009) or in food packaging. Assessing exposures to ENMs poses challenges because of the need to characterize and quantify the material once it is released and to assess its stability and potential biotransformation during food processing or in food (FAO/WHO 2009). Critical research questions include, ***What is the propensity of ENMs to survive in the gastrointestinal tract, particularly the acidic gastric milieu, as particles? If they survive, what is the extent of absorption and assimilation into the organism?***

Once nanomaterials enter the human body, their surfaces may be modified by native biomolecules, and these modifications may influence their dosimetry. The process, referred to as opsonization or differential adsorption, involves adsorption of proteins and lipids onto the surface of ENMs (protein corona formation), which potentially modifies their size, surface charge, and aggregation state (Muller and Keck 2004; Lynch et al. 2007). Research focused on biologic surface modification is increasing, but the topic is complex. Such fundamental nanoparticle characteristics as hydrophobicity, size, and charge probably dictate the composition of the corona[5] of a nanoparticle. Moreover, the dynamics of biomolecular association are not always on the same timescale. Recent evidence indicates the formation of a "hard" corona with stable proteins and an outer, "weaker" corona that has quickly exchanging proteins (Walczyk et al. 2010;

---

[4]It should be noted that defining "persistence" of an ENM is more challenging than for traditional molecules with defined molecular formulas. This is true for all processes that alter the form of the ENM from its pristine state to a transformed one. Chapter 4 provides additional discussion on the appropriate metrics for defining alteration or degradation rates.

[5]The corona is the coating of proteins that bind to the surface when nanoparticles interact with biologic fluids.

Monopoli et al. 2011). ***A critical research question is, What are the nature and implications of biomolecular modifications of ENMs?***

Environment—Needed Research on Modifications and Exposure Assessment

As discussed earlier, modification of ENMs in the environment is a key element of their exposure potential. Publications over the last 5 years have highlighted the diversity and complexity of nanomaterial modifications. ENMs can dissolve, aggregate, disaggregate, agglomerate, disagglomerate, or be chemically transformed in environmental systems (for example, sulfidation or adsorption of Natural Organic Matter (NOM)). Specifically in the atmosphere, released ENMs may become incorporated into preexisting atmospheric particles, or may be coated through adsorption or condensation of atmospheric vapors.

A given ENM, if discharged to a stream, could have a physical and chemical composition and a fate different from what would follow application to plants in fertilizer in an agricultural field. There is a need to understand the transformation processes and their variation with ENM structure. Adding to the research challenge is the fact that these processes (for example, ENM aggregation) can affect transport and fate, exposure, and ultimately toxicity. Research approaches need to recognize the complexity of the underlying processes and use systems approaches to examine the interdependencies of the processes. The models discussed in Chapter 4 are essential tools for addressing these challenges.

Of all the ENM modification processes, aggregation (both homoaggregation and heteroaggregation) is the most central to environmental health (Hotze et al. 2010). Aggregation is a result of the attachment of particles to themselves (homoaggregation) or to other environmental surfaces (heteroaggregation). The attachment of ENMs to surfaces depends heavily on the solution conditions (for example, pH, ionic strength, and ionic composition) and on the physics of the attachment, that is, how the ENM approaches the surfaces of particles (Mylon et al. 2004; Hotze et al. 2010). The presence of organic matter or biomacromolecules also substantially affects an ENM's attachment to surfaces (Wiesner et al. 2009; Phenrat et al. 2010; Saleh et al. 2010). If nanomaterials collect into larger micrometer-size aggregates, their transport and reactivity may be different from those of materials that remain as isolated nanoscale high surface-to-volume materials.

Aggregation into larger particles may also affect interactions with receptors. ENMs may interact with other particles in the environment, such as clay particles (heteroaggregation) that can affect their transport and distribution. For example, heteroaggregation of ENMs with soil particles will alter their transport to water by runoff (for example, from croplands that are applied with biosolids). Heteroaggregation of ENMs with larger airborne or waterborne particles will increase their rate of deposition from air or their sedimentation rate in water, respectively. Sedimentation of ENMs out of the water column will decrease water column ENM concentrations and increase sediment ENM concentrations (Wiesner et al. 2009; Hotze et al. 2010). These processes will affect the organisms that are likely to be exposed, the exposure routes, and the effects of exposure. There is also evidence that nanoparticle aggregates may remain stable in suspension and maintain toxicity potential despite being present in an aggregated state. For example, Salonen et al. (2008) found that C70 fullerenes formed "stable, homogeneous suspensions" in water through interaction with phenolic acids that are present in and released from plant matter, and Fortner et al. (2005) identified the formation of stable suspensions of "nanocrystals" (25-500 nm diameter) of C60 fullerene aggregates in water. Fortner et al. (2005) also found that these aggregated fullerene nanocrystals exhibited antimicrobial activity, suppressing bacterial growth and respiration. Lyon and Alvarez (2008) also cited a number of studies demonstrating that these nanoscale aggregates in water can yield a material with toxicity to aquatic invertebrates, fish, and the cells of higher organisms, and that the aggregates can enter and accumulate in cells. Finally, Salonen et al. (2008) showed that phenolic acid-coated C70 aggregates could translocate across the membranes of human cells in culture. In addition, they induced the contraction and death of those cells through agglomeration and aggregation into micro-sized particles that interacted with the cell membranes. ***A detailed understanding of ENM aggregation is needed to create models of fate and transport of ENMs in the environment.***

A complicating factor for aggregation and heteroaggregation is the presence of polymer or surfactant coatings used to stabilize ENMs against aggregation sterically or electrosterically in the absence of sufficient charge stabilization. Macromolecules attached to ENM surfaces greatly affect their attachment behavior, including attachment to NOM (Saleh al. 2008; Petosa et al. 2010). Nearly all ENMs in the environment are expected to have an engineered macromolecular coating or will become coated with NOM.

Research findings on the effects of coatings or of adsorbed NOM on the transport of ENMs in the environment are limited and contradictory. In some cases, the coatings have been shown to prevent aggregation; in others, they have been shown to increase aggregation (Jarvie et al. 2009). Even though the dual role of NOM on the aggregation (flocculation) and dispersion of colloids has been studied by many different scientific disciplines, the complexity of NOM, and the very small size of ENMs compared with the size of the adsorbed macromolecules, complicates predictions of the effects of NOM on polymer-coated ENMs.

The fate of the engineered coatings in the environment is not well established. Once discharged into environmental waters, the engineered coatings may be removed, and this change can cause nanomaterials to aggregate. However, some covalently bound polymeric coatings may be resistant to removal or biodegradation and remain on the ENMs. Similarly, NOM can coat ENM surfaces and act as a natural surfactant, preventing aggregation or promoting disaggregation. The ability of a coating to promote or prevent aggregation will probably depend on the ENM surface, the coating properties, and the environmental milieu. ***There is a notable gap in information about the coatings on ENMs and how their presence and stability are related to ENM aggregation and ultimate fate in the environment.***

Another important issue in assessing exposure to ENMs is nanomaterial transformation and persistence in the environment. ENM transformations in the environment may lead to materials that have different partitioning, transport, and toxicity characteristics from the native ENM. Chemical transformations in the environment can include dissolution, sulfidation, oxidation-reduction, photodegradation, biodegradation, adsorption of organic matter and biomacromolecules, and biodegradation of macromolecular coatings. Biologic oxidation or reduction and biodegradation of macromolecular coatings will alter the surface properties of ENMs and therefore their transport and distribution in the environment. The oxidation of zero valent iron nanoparticles or the adsorption of NOM has been shown to increase their mobility in porous media (Phenrat et al. 2009a) and decrease their toxicity to bacteria (Auffan et al. 2008; Li et al. 2010) and mammalian cells (Phenrat et al. 2009b). The dissolution of silver nanoparticles correlates with their toxicity to bacteria (Bae et al. 2010). Sulfidation of silver nanoparticles in the environment may decrease the release of silver ions and therefore their toxicity (Liu et al. 2010; Levard et al. 2011).

How those chemical transformations (such as dissolution) affect the persistence of ENMs in the environment remains unknown. Zinc oxide and silver nanoparticles are two examples in which dissolution affects persistence, but this phenomenon occurs with many types of ENMs. The persistence of organic and fullerene systems is a function of their redox reactivity; for example, fullerenes in water can be oxidized easily in the presence of light (Hou et al. 2010). It is important to note that the transformations do not necessarily operate singly or in series. It is unclear how sulfidation or aggregation of silver nanoparticles affects their rate of dissolution and persistence in the environment. Although the simple reaction chemistry of many of the most common ENMs is established, ***the quantitation of the dissolution rate and dependence on environmental conditions remains a critical research gap.*** Only with this information can the persistence of ENMs be clearly defined.

The transport and fate of ENMs are coupled; that is, modifications will affect their transport and ultimately affect their fate. Once ENM modifications are understood, their transport in the environment can be considered. The transport of ENMs in the environment will determine their probable accumulation points, potential exposure routes, and where dilution occurs. The latter information is needed to predict ENM concentrations in environmental media. There is little knowledge about the transport and distribution of ENMs in the environment after release.

Transport in porous media is an important mechanism to consider. Whereas aggregation is attachment of ENMs to other ENMs or to suspended particles, deposition involves attachment of ENMs to fixed porous media, such as soil, sediment, or filter media used in water treatment. Strong attachment to porous media suggests minimal transport in the aquatic environment and removal in drinking-water treatment systems. However, attachment of ENMs to wastewater solids presents additional pathways for environmental exposures.

Understanding deposition is an important precursor to accurate exposure assessment, in that studies of deposition can provide information on environmental sinks for ENMs. For example, exposure modeling for nanoscale titania that relied on bulk attachment behavior suggested that sludge is the likely sink for these ENMs (Gottschalk et al. 2009); this information can be used to determine the distribution of ENMs in the environment and the likelihood that existing control strategies (such as the use of activated carbon or sand filters in water treatment) can mitigate ENM exposure.

A continuing research theme is the systematic linkage between ENM properties and their deposition and transport behavior in model and real porous media. As for aggregation, an ***improved understanding of the properties of ENMs that affect attachment to porous media will allow better prediction of deposition in environmental media.***

Factors that influence aggregation of ENMs and their attachment to porous media will be useful in assessing the distribution of ENMs in the environment. However, tools and methods are needed to measure the occurrence of ENMs at low concentrations in environmental media. In contrast with human health exposure assessment, monitoring for environmental exposure to ENMs is in its infancy. Several studies have used microscopy to examine sludge and sediment to locate ENMs (Kim et al. 2010). However, it is unlikely that such an approach would be scalable or routine. Other potential tools for environmental monitoring are difficult to use because of interference from naturally occurring nanomaterials, the low concentration of ENMs in environmental samples, and the inability to detect individual and transformed ENMs. As discussed above, tools for conducting real-time monitoring of ENMs are needed.

Ultimately an environmental exposure-assessment model that contains important ENM transformations as subcomponents is needed. However, a 2010 state-of-the-science report (J.M. Johnston et al. 2010) concluded that there are many data gaps in environmental-exposure models; for example, "problem formulation" is inadequate for assessing environmental and ecologic exposures to ENMs, and there is a need to validate and assign values to parameters for the models (screening or otherwise). Those gaps arise from the particulate nature of ENMs and the absence of data that are needed to properly assess appropriate parameters for the models. For example, needed parameters include assimilation efficiency; rates of emission of ENMs to the environment; ENM properties that affect transport in air, porous media, and water columns; interphase mass transfer, such as runoff from land to water; degradation rates; dilution rates; sedimentation rates; and distribution coefficients between phases. One exception to the need for parameters, may be the ability of these models to track particulate matter from an airborne source (such as a smokestack) to receptors on the basis of data on rates of dry and wet deposition of airborne particulate matter.[6] ***How can the fate and transport of ENMs be fully described and modeled?*** There are two approaches to model the fate of ENMs in the environment. One approach uses measured bulk parameters (such as distribution coefficients) in applied empirical, deterministic, or probabilistic models for heterogeneous and large-scale systems. A second approach builds mechanistic models based on fundamental processes affecting the behavior of ENMs in natural systems. Identifying the relevant processes should be possible in accordance with principles of colloidal science and an understanding, albeit limited, of the behavior of ENMs in environmental media. It is critical to identify those conditions (for example, size or other properties) under which ENMs can be modeled like colloids, and those conditions that allow them to be modeled like chemicals. It remains unclear if mechanistic models can be scaled up to the ecosystem level. Models capable of estimating ENM distribution in the environment, combined with an understanding of the sources of environmental ENMs (discussed below), will allow research to address questions that provide the greatest potential to reduce uncertainty in environmental exposure-assessment models. Source information may include descriptions of runoff from agriculture due to ENMs in biosolids applied to the fields; processes of ENM dry and wet deposition from air, sedimentation, and later sediment transport; and groundwater infiltration from agriculture activities.

Quantifiable Dose, Biodistribution, and Bioaccumulation

When a material interacts with a biologic receptor, hazard assessment requires the definition of the quantifiable dose of the material. In the case of ENMs, the connection between the amount of material at the interface of an organism and its relevant bioavailability, which can be different from the dose, is largely unknown. For example, it is not clear whether the material mass, surface area, or number concentration is the most appropriate metric for assessing the dose of nanomaterials, since the relationship among the dose metrics may change as the nanomaterials interact with biologic receptors. Physical and chemical modifications of ENMs can have a substantial effect on their bioavailability and biodistribution. The dose of an ENM depends on its distribution in the compartments of an organism; distribution data can be essential for defining hazards to specific organs and tissues. Thus, the fundamental metric of concentration in an organism is not necessarily the best measure of dose.

---

[6]However, the particulate nature of ENMs may contribute to behaviors that will differ from that of chemicals. This is true of the parameter describing the distribution of ENMs between phases. The distribution of a molecule (for example, a dioxin) is based on the equilibrium partitioning of that molecule between two phases and can be determined from thermodynamic constants. Conversely, the distribution of ENMs will likely result from the kinetics of attachment to other particles or environmental surfaces.

Human Health—Biodistribution and Dosimetry

Information about the biodistribution of ENMs after exposure by inhalation or oral or dermal uptake is essential for the determination of relevant doses, particularly for designing in vitro studies and interpreting their findings. Similarly, doses used in biokinetic animal studies need to be informed by relevant data on human exposure, whether in a workplace, in a laboratory, or in consumer use of nano-enabled products. Those data are critical to guide the dosimetry of in vitro studies. If only a tiny fraction of inhaled nanoparticles can be expected to reach the brain (say, much less than 0.1%) (Semmler-Behnke et al. 2008; Kreyling et al. 2009), what concentration of nanoparticles would be appropriate to use in an in vitro study involving exposure of neuronal cells?

An outstanding ENM-dosimetry question is, ***What is the most appropriate metric for assessing ENM dose?*** This is a complex issue that is peculiar to the study of nanoparticles. ENMs are chemical objects whose molecular weights are hundreds or thousands of times greater than those of most molecules; moreover, particle mass is size-dependent. For example, a parts-per-billion suspension of metal nanoparticles that are 5 nm in diameter has 25 times the surface area of a batch of 25-nm nanoparticles of the same metal and 125 times the number concentration. Whether the dose should be expressed as particle mass, particle surface area, or number depends on the objective of the study. For example, data from inhalation and instillation studies suggest that at least for some toxicologic end points surface area is the appropriate metric for gauging effects. For well-characterized ENMs, it is straightforward to express dose in various ways, and dose issues can be explored with appropriate study designs.

The characteristics of the ENM sample may influence the metric used. For example, measuring airborne nanoparticles at very low mass concentration would most reliably use number concentration, whereas concentrations in tissue samples may be based on a chemical analysis of mass or possibly transmission electron microscopy of number concentration. Mechanistic studies that explore what is an appropriate metric for expressing ENM dose in organisms, tissues, or cells (microdosimetry) are needed. Specifically, there is a need to define and determine biologically or toxicologically based metrics for dose, such as biologically available surface area or surface reactivity, recognizing that the appropriate metric will depend on its intended purpose and underlying mechanisms.

When designing animal studies, researchers are challenged in extrapolating findings to real-world human exposure to ENMs. For inhalation studies, this includes not only ensuring that the physical form of the aerosol (for example, agglomeration state and particle size distribution) is similar to the form in the environment, but consideration also is needed of the differences between humans and experimental animals in deposition efficiency throughout the respiratory tract to adjust for breathing mode, airway geometry, and associated inhalability and respirability of the particles in question. For example, aerodynemically larger particles (that is agglomerated or aggregated ENMs) may be respirable by humans but not respirable by mice or rats.

Several models have been developed to predict deposition efficiencies of inhaled isometric particles in the human respiratory tract, most notably the International Commission on Radiological Protection model (ICRP 1994) and the Multiple Pass Particle Dosimetry (MPPD) model (Asgharian et al. 1999). The MPPD model has been expanded to estimate the deposited fraction of airborne isometric particles in rats, which is useful for dosimetric extrapolation of results from rat inhalation studies to humans. However, although those models simulate different breathing scenarios (resting, exercising, and working strenuously), inhalability, and diverse particle (including nanomaterial) characteristics (size distribution, density, and concentration), they are not useful for modeling the effects of different particle shapes. Thus, given the multitude of nanoparticle shapes—for example, fibers, tubes, aggregates, and agglomerates—there is a need to develop deposition models for human and rodent respiratory tracts that can be validated experimentally. In addition, although the MPPD model allows one to model pulmonary retention and accumulation, an expansion is needed to include particle shape and translocation from the deposition site in the respiratory tract to other organs. ***Two critical research needs are to refine inhalation exposure and deposition models and to develop similar models for ingestion and dermal exposure.***

Virtually all analyses of research gaps in this regard have highlighted the importance of validating and linking in vitro and in vivo studies of ENMs. The models mentioned above provide one way to address an issue that is related to the scaled dose in different organisms. The ICRP model is restricted to the human respiratory tract, but the MPPD model is applicable to both the rat and the human respiratory tracts. Once sufficient data have been collected on exposure and hazards associated with a specific ENM, these predictive deposition models will permit the dosimetric extrapolation of toxicologic information obtained from acute or longer-term rodent studies to establish exposure limits and risk estimates for humans. Figure 3-5 outlines this concept of dosimetric extrapolation as it may be applied to defining the human-equivalent concentration (HEC)[7] for inhalation exposure. The deposited doses can be considered for short-term effects (short-term HEC) and for chronic effects (long-term HEC); with information on both rodent and human retention data, accumulated doses can be considered. Moreover, if available databases on cell types and their numbers in a specific lung region are used, the deposited dose per cell (microdosimetry) can be estimated. ***Research that validates dosimetric extrapolations of exposure between different kinds of organisms is critical and will ultimately place high-throughput in vitro studies in the appropriate context.***

Following uptake, ENMs may distribute throughout an organism and reside in locations distal to the initial exposure site. Nanomedicine publications offer insights into ENM biodistribution in mammals, particularly on the basis of rodent models and several routes of entry. Typically, particles larger than 10 nm

---

[7]The HEC is the quantity of an agent that, when administered to humans, produces an effect equal to that produced in test animals with a specified smaller dose.

in diameter that are administered intravenously or intraperitoneally are found in the liver; however, circulation times in the blood depend heavily on the surface properties of the materials. Subdermal injection led to appearance of ENMs in the lymph system (Ohl et al. 2004; Gopee et al. 2007; Moghimi and Moghimi 2008). After clearance into the bloodstream, one elimination pathway is probably in feces via hepato-biliary elimination (Kolosnjaj-Tabi et al. 2010). Clearance through urine of isometric ENMs up to 9 nm in diameter has been observed, and urinary clearance of materials of high aspect ratio and even greater length has been reported (Choi et al. 2007). Those and other studies have provided qualitative guidance on the likely target organs for ENM exposure, but they also have highlighted the role of surface modifications and routes of entry for distribution to the final organ.

Knowledge of the biodistribution of nanoparticles after inhalation, oral, or dermal uptake is essential for identifying specific organs that may be targeted, including injury mechanisms, and designing the toxicity assays that best represent the exposures and mechanisms of toxicity. That aspect of nanotechnology-related EHS research has not been extensively explored, in part because labels for tracking ENMs require additional development to ensure their stability in vivo.

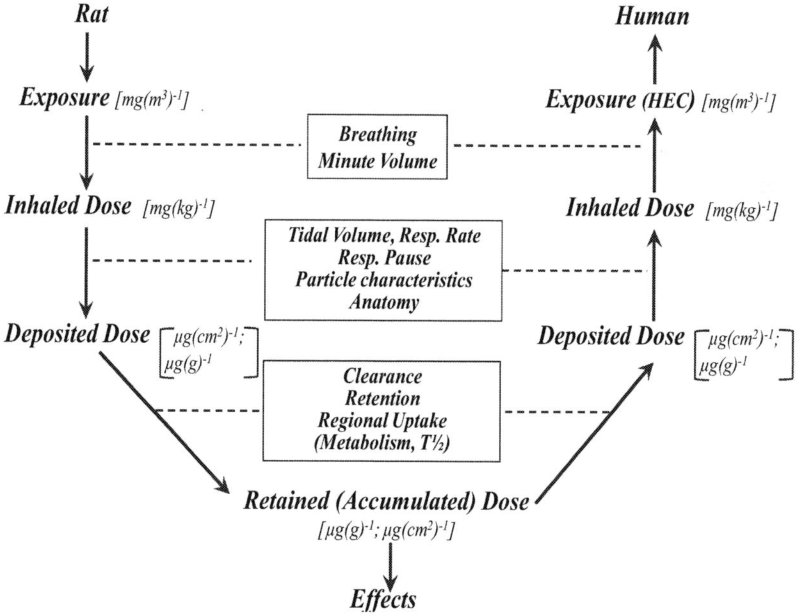

FIGURE 3-5 Extrapolation of dosimetry of inhaled particles from rats to humans. The assumption is made that if retained dose is the same in rats and humans, then the effects will be the same. Source: Adapted from Oberdörster 1989.

Elimination is another process for which data and valid models are needed. Nanomaterials may display long retention half-times in some organs, but few studies (Schluep et al. 2009; Pauluhn 2010; Maldiney et al. 2011) have quantified or even determined elimination routes (Bazile et al. 1992; Alexis et al. 2008). Although fecal excretion (hepatobiliary pathway) and urinary excretion (of particles smaller than 5-9 nm) have been described (Choi et al. 2007; Lacerda et al. 2008a,b; Kolosnjaj-Tabi et al. 2010), quantitative elimination models need to be developed. For example, what is the elimination pathway of nanomaterials that accumulate in the central nervous system by bypassing the blood-brain barrier through neuronal translocation of inhaled nanoparticles from nasal deposits to the olfactory bulb (Oberdörster et al. 2004)?

What are the most valid quantitative biokinetic models that relate ENM properties to their distribution in organisms, organs, and tissues? A crucial component of this research will be the development of tools that enable the detection of ENMs or their constituents in tissues for understanding the biokinetics of ENMs. Studies of biokinetics should be integrated with evaluations of ENM modifications. Establishing a comprehensive biodistribution model—including uptake, translocation, and elimination pathways and mechanisms—will be an important input for bioinformatics.

Biodistribution and Bioaccumulation in the Environment

ENMs will persist or accumulate mainly in the solid and aqueous phases of the environment, unless they are suspended in the atmosphere. Such environmental media may act as diluting agents only if the ENMs do not preferentially distribute into specific environmental (for example, sediment or air-water interface) or biologic (for example, gills) compartments. Lessons from other low-concentration molecular contaminants (for example, methyl mercury) reveal that processes can concentrate materials in specific compartments and thereby increase their relative dose and possible effects. Many aspects of distribution are physicochemically based, but in environmental systems biologic compartments are important sinks. Although many issues in this discussion are relevant to both ENM sources and ENM transformations, they are discussed here because of their relevance to biologic settings.

There is a need to understand the potential for ENMs to accumulate in particular environmental compartments and to determine the area over which ENMs are distributed. An understanding of this distribution is needed to address dilution potential of ENMs released from a point source (for example, wastewater-treatment plant effluent discharge or exhausts to air) and will help to focus risk assessment of ENMs on the relevant environmental compartments. ENM distribution in the environment will be controlled largely by attachment to other ENMs or to environmental surfaces, such as minerals, plant leaves, and fish gills. Strong attachment to surfaces affects aggregation and sedimentation in aqueous environments and possibly bioavailability, bioconcentration, and persistence in the environment.

Attachment to biosolids, which will also affect the sources of environmental ENMs, is a critical issue to address. For example, recent evidence suggests that certain ENMs are leaving wastewater-treatment plants in biosolids and effluent water (Kiser et al. 2009; Kim et al. 2010). A screening-level exposure model for ENMs will require estimates of the distribution of ENMs between biosolids in a wastewater-treatment plant and the effluent water going from the plant to a receiving body of water. Strong attachment of ENMs to biosolids may suggest that terrestrial exposure from biosolids that are spread on croplands is a greater source of ENMs than aquatic exposure to ENMs from wastewater-treatment plant effluent. Such distribution data can also be used to allocate ENM sources to their appropriate environmental compartments better, but this requires an ability to measure and characterize ENMs in complex environmental matrices—a research issue highlighted in Chapter 4. That information can be enhanced by developing an understanding of the ENM properties that most affect attachment of ENMs to environmentally relevant surfaces (for example, biosolids, clays, and cell walls). Ultimately, distribution coefficients of ENMs can be estimated by using a small set of ENM characteristics. The distribution coefficients can then be incorporated into screening-level exposure models on a material-by-material basis to decrease uncertainty in the exposure modeling. Validation of models that predict the distribution of ENMs between phases will require measuring ENMs in complex natural media (soil, sediment, and air).

Assuming that the models that are created to determine sources, transport, and transformations of ENMs can provide reasonable estimates of the concentration and physicochemical form of ENMs in particular environmental compartments, the fractions of the transformed ENMs that are bioavailable to target receptors need to be determined. This includes bioconcentration of ENMs themselves, bioavailability of toxic metals released from ENMs, and uptake of toxins associated with the ENMs. ***Indeed, the bioavailability of ENMs to organisms is poorly understood, and a better understanding of it would enhance ecotoxicologic studies.***

It is critical to address questions about ENM bioaccumulation. Specifically, can ENMs bioaccumulate? If they can, to what extent, and what specific properties are most critical for bioaccumulation? Models for predicting bioaccumulation of contaminants in fish and fish populations are available, but not similar models for sediment-dwelling organisms and other sensitive aquatic species in the water column, such as Daphnids or aquatic plants that may be found to be sensitive receptors on the basis of the risk assessment. The biologic and physical processes that affect bioconcentration and trophic transfer of ENMs will be very different from those of molecular contaminants. For example, nanoparticle size, aggregate size, coating properties, and aspect ratio may influence bioavailability and uptake. The influence of those properties on biologic uptake is not known. The presence of soil has also been shown to affect the toxicity of certain ENMs greatly; for example, fullerenes had little influence on soil bacteria, because of the attachment of the fullerenes in the soil organic matter, in comparison with their influence on bacteria in aqueous solutions (Tong et al.

2007). In contrast, nanosilver in wastewater biosolids applied to soil was found to inhibit plant growth and reduce soil microbial biomass (Zeliadt 2010). Therefore, bioavailability and uptake measurements need to include environmentally relevant surfaces (for example, soil for terrestrial uptake and suspended solids and organic colloids for aquatic organisms). New models for bioavailability and uptake are probably needed for ENMs and will need to be species-specific.

Organism and Ecosystem Effects of Engineered Nanomaterials

The responses of humans, other organisms, and the larger ecosystem to ENMs are central to understanding potential risks. Hazard assessments involving single organisms or in vitro toxicity assays have been the focus of research in this field, and there have been many in vitro and in vivo studies of various cell lines and organisms. Most studies use a single material; however, there is incomplete information on effects of the array of nanomaterials currently used in products, in part because they are not available to researchers. Many studies address effects of acute exposure, but there is a lack of information on effects of chronic exposure (involving time-course experiments). However, data from acute and subchronic studies can be very valuable in selecting doses and determining end points for chronic studies. In addition, doses in acute studies are high relative to likely "real world" exposure, and there is variability in how nanomaterials are introduced into exposure suspensions and therefore uncertainty regarding how such study conditions may influence effects. There is also uncertainty regarding how the media or biologic fluids interact with the ENMs to alter their effects (see section on Modifications of and Exposures to Engineered Nanomaterials). Finally, there is little information on the effects of nanomaterials on populations and communities of organisms.

Human Health Effects

Human-health hazard identification has been conducted for several ENMs using in vitro and in vivo methods (Cui et al. 2005; Oberdorster et al. 2007; Lewinski et al. 2008), and many studies have indicated a relationship between dose (often at extremely high doses with questionable relevance to human doses) to ENMs and a toxic response. Fewer studies have addressed dose-response issues (Wittmaack 2007; H.J. Johnston et al. 2010). Human exposure models and measurement tools are also available for assessing human exposure to ENMs in the workplace. Despite availability of those models and tools, however, few exposure (largely workplace) studies are available and there is a lack of readily available measurement methods—a gap that needs to be closed in the short term. Such research is critical in that it permits stakeholders to gauge risk on the basis of actual, relevant exposures and doses.

Because most hazard assessments have relied on in vitro testing with doses that tend to be higher than realistic exposures an important question is,

What biologic effects occur at realistic ENM doses and dose rates, and how do ENM properties influence the magnitude of these effects?

The ability of in vitro high-throughput testing to provide information on what happens in vivo has not been demonstrated, and proper in vitro tests have not been developed to examine the numerous species. A conceptual approach for toxicity testing of nanomaterials that begins to address that matter is illustrated in Figure 3-6. In addition to several in vitro and in vivo components, the figure emphasizes the need for providing exposure and hazard data that are essential for risk assessment. There is a need to characterize the relationships between in vitro and in vivo responses. Studies directed at these relationships will require standardized and validated in vitro methods (for example standardized cell types and exposure protocols) that represent specific exposure routes and validation of results from in vitro studies with responses from relevant in vivo studies. This research is vital for developing high-throughput screening strategies for ENMs. ***A long-term goal is to develop simple in vitro assays that predict in vivo effects at the organism level and may eventually be used for high-throughput screening assays.***

A key requirement should be that any in vitro assay used as a predictive tool needs to have been validated with appropriate in vivo studies. Other considerations include the following:

- Results of simple assays only identify a hazard and should be used solely for ranking, for example, to establish a hazard scale (Rushton et al. 2010).
- Mechanistic pathways discovered in in vitro studies based on extraordinarily high unrealistic doses probably do not operate in vivo in real-world conditions, because mechanisms are dose-dependent (Slikker et al. 2004a,b), and should be interpreted with caution.
- In vitro results reflect acute responses and could be highly misleading in predicting long-term effects. For example, soluble zinc oxide nanoparticles induce substantial oxidative stress responses that do not persist, because of solubility and induction of effective adaptive responses. Drinker et al. (1927) described this in workers exposed to zinc fumes.

Another important research gap is the underlying biology of ENM interactions. Some nanomaterial toxicity mechanisms have been investigated, mostly under limited study conditions (for example, dose and time courses described above). Several toxicity mechanisms are described in the literature on inhalation-particle toxicology and related diseases, including inflammation and oxidative stress, immunologic effects, protein aggregation and misfolding, and DNA damage. The previous research on ENM effects has been focused largely on inflammation. However, more information on those and other mechanisms is needed. For each of the mechanisms, the characteristics of the ENM (such as size, surface properties, and composition) that are associated with a particular biologic effect should be identified along with the specific effects. ***How do ENM properties influence toxicologic mechanisms of action?***

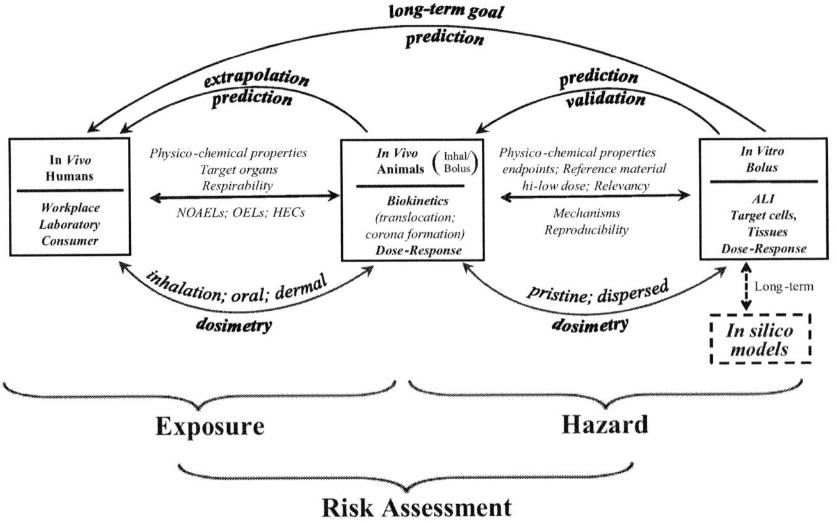

FIGURE 3-6 Concept of ENM toxicity testing for human health risk assessment. Risk assessment requires information on hazard, exposure, and exposure–response (or dose–response) relationships. Hazard identification and hazard characterization (left and middle boxes) may use in vitro and in vivo methods. In vitro and in vivo studies involve many considerations, including the physical–chemical properties of the nanomaterials, the method of administering the nanomaterials, the target cells or tissues, and the dose–response relationship. Comparison of the correlations between in vitro and in vivo responses—with in vivo data as the standard—is needed. The lower, bidirectional arrows refer to the dosimetric correlations between in vitro/in vivo animals and in vivo animals/in vivo humans with the goal of informing the design of in vivo animal studies by using available human exposure data and dose–response information from animal studies to compare with human data. The upper, unidirectional arrow refers to extrapolating effects and mechanisms from relevant animal studies to humans with the goal of deriving recommended exposure limits (OELs). In the long term, in silico models may be developed to assess hazard. Abbreviations: ALI, air–liquid interface; NOAEL, no-observable-adverse-effect level; OEL, occupational exposure limit; HEC, human-equivalent concentration. Source: Adapted from Oberdörster 2011.

Ecologic Effects

Although there are many gaps in our understanding of potential human health risks from ENMs, the gaps in our understanding of potential ecologic risks are considerably greater (Bernhardt et al. 2010). One reason is that environmental hazards and exposures are more complicated owing to the greater number of potential exposure routes and receptors and the complex relationships among organism effects, population effects, and ecosystem responses. The complexity makes it difficult to define the problem that is being addressed in an ecologic risk assessment—the potential organisms affected and the ecologic effects.

Proper problem formulation (as discussed in Chapter 2) is a critical first step in risk assessment (EPA 1998; NRC 2009) and has yet to be adequately considered for ENMs, because potentially affected organisms and effects on ecosystem function depend on the release points, fate, and transport of ENMs in the environment; these factors are unknown at present. The process of problem formulation for ecosystem response to ENMs will require a tiered approach, given that the ecologic end points are not known and the relationships between organism, population, and ecosystem responses are poorly understood (Bernhardt et al. 2010). Toxicity modeling and testing will benefit from models and measurements on fate and transport of ENMs as this will help to determine concentrations that will reach ecologic targets.

Like the goal of human hazard assessment, the goal of ecologic hazard assessment is to predict the potential for toxicity to organisms, communities of organisms, and ecosystem processes with relevant assays (Figure 3-7). Determining the potential ecologic impact of nanomaterials is challenging, given the various types of organisms found in different environments, their various life-history characteristics, and their differing physiology. Research is needed to guide selection of appropriate ecologic receptors, to develop appropriate ENM assays, and to conduct model ecosystem studies that address potential effects on a larger scale, such as the population, community, or ecosystem.

Gaps to be addressed include characterization of low-dose effects, assessment of multiple end points over a life cycle, and research on effects on multiple organisms along various pathways. Information on the actual pathways that are disrupted in whole-organism assays is critical. Duration of exposure should be considered in relation to effects, inasmuch as effects may change owing to accumulation or formation of byproducts in the organism or recovery pathways.

In general, all current ecologic testing strategies use single organisms, and effects are predicted from one or two model species. Data will be needed to predict sensitive species and higher-order effects on communities and ecosystems, including interactions among species, species community assemblages, biodiversity, and ecosystem function. The inability to predict such effects creates great uncertainty regarding the potential effects of ENMs on the ecosystem.

Gaps in Data on Ecologic Effects of Engineered Nanomaterials

Numerous standard screening-level toxicity tests for specific aquatic and terrestrial organisms have been proposed for evaluating the effects of ENMs. However, several data gaps need to be addressed to ensure that the tests can predict ecosystem impacts of ENMs. The first is a poor understanding of the mechanisms of toxicity. Most ecologic effects have focused on LC_{50} data; only a few studies have examined specific effects or mechanisms by which nanomaterials act on organisms. Lethality is time-dependent and chemical-dependent, and using LC_{50} data introduces bias into modeling through use of artificial periods

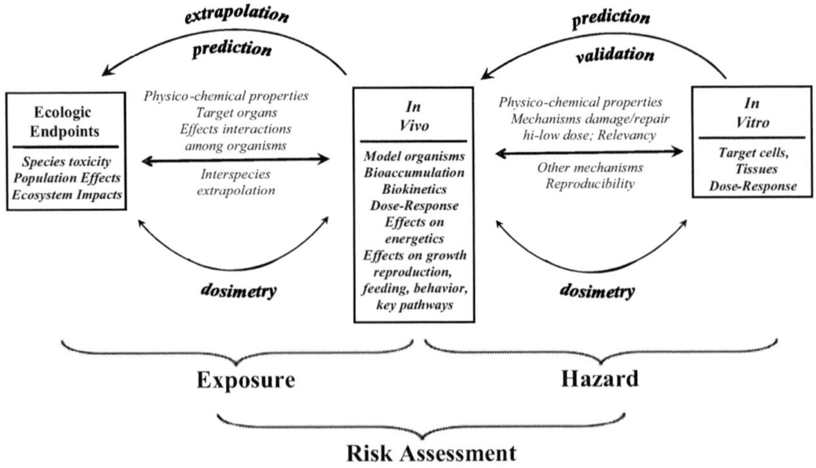

FIGURE 3-7 Ecologic hazard end points for making predictions of the environmental effects of nanomaterials. As set out in Figure 3-6, to assess the environmental risks from nanomaterials, information on hazard, exposure, and exposure-response is needed. In vitro and in vivo assays are used for assessing hazard. Conducting in vitro and in vivo studies involves considerations of the physical-chemical properties of the nanomaterials, dose-response relationships, mechanisms of toxicity, and relevancy of the tests for providing useful measures, including understanding of effects on larger organisms and populations. Extrapolating from in vivo effects to ecologic end points will benefit from exposure measurements and models to understand concentrations that will reach ecologic end points. Source: Adapted from Oberdörster 2011.

that do not take intermediate end points into account (Heckmann et al. 2010; Jager et al. 2010). The ability to predict toxicity on the basis of the properties of ENMs will require some knowledge of toxicity mechanisms, in addition to data on acute and gross end points, such as death. As mentioned in the case of human toxicology models, oxidative stress is a toxicity end point that is being explored (Nel et al. 2006; Xia et al. 2006), but there have been few studies of it in ecologic models (Klaper et al. 2009). Leveraging research advances to include key ecologic receptors may help to correlate ENM properties with their potential for ecologic damage. However, questions regarding the oxidative-stress response should be addressed: How much oxidative stress is "dangerous," and when? How should response changes over time be interpreted; for example, is there a short-term reaction to the ENMs that has no long-term effects? Is oxidative stress the best end point to monitor for nanomaterial effects? ***Thus, future ecotoxicologic research should focus on improving understanding of toxicity mechanisms of ENMs.***

More information is needed on the pathways of biochemical responses to ENMs and their various properties and on the linking of the responses to adverse outcomes. In particular, pathways of adverse outcomes of exposure to nanomaterials need to be defined beyond those of oxidative stress. Pathways of adverse outcomes (for example, effects on survival and reproduction) need to be determined; that is, changes in the biologic mechanisms need to be linked to larger organism or population outcomes to yield useful measures. Researchers must evaluate how a biologic property (such as oxidative stress) translates to effects on organism survival and reproduction. Adaptation may occur after repeated exposures and make organisms tolerant to higher or longer exposures. Many other toxicity pathways might provide better and more sensitive information on ENM effects on multiple receptors. Another consideration is the dose of the nanomaterial used in molecular assays. Specifically, what dose is appropriate for investigating what pathways?

A research issue related to organism testing is the effects of ENMs on other end points (aside from death) and after low-dose chronic exposure. Because ENMs will probably exist at very low concentrations in the environment and will persist, low-level chronic exposure is the most likely scenario. Thus, tests with chronic exposure should be developed and validated for ENMs and other end points, including effects on growth, reproduction, metabolism, and behavior; and these effects need to be considered in testing strategies. Low-level chronic exposure studies pose challenges for nanomaterial dosing. For example, should ENMs be reintroduced as they settle out of an assay? What methods should be used for dispersing particles into the media? How can exposure and changes in ENMs be monitored in the assay?

ENMs may have a variety of effects at the population level, such as effects on population dynamics, reproduction, genetic structure, demography, and ultimately the sustainability of a population. Therefore, it is important to determine ***how ENMs and their properties affect populations.***

For key members of systems, some effects may be measured by using chronic-assay end points mentioned above (such as effects on reproduction or growth), and measurements of age, class structure, and population genetics may provide population-level information. Examining changes in the population genetics or age structure of a population is not a standard approach, but such basic studies will be necessary to determine how a population of organisms will behave when in contact with ENMs—often the most important indicator from the perspective of ecologic risk assessment and sustainability. The death or injury of an individual fish or small group of fish (organism-level response) may not be cause for alarm, but the crash of an entire population may create great problems not only for that species but for the larger community or ecosystem.

ENMs may have more than overt acute toxic effects on specific species. Depending on their properties, ENMs may alter interactions among organisms in a community, for example, by changing predation, commensalism, or dominance. Such effects are important to understand on a large scale. The few data that exist suggest that, at a minimum, nanomaterial exposure may affect bacterial community structure (Lyon et al. 2008). In addition, ENMs may influence such ecosystem functions as nutrient cycling, energy, productivity, and biomass by affecting communities or organisms that are critical for these functions, for example, changing the abundance of nitrifying bacteria or the availability of nutrients. The questions remain: ***How do the properties of ENMs influence their impact on community structure? How do the properties of ENMs influence ecosystem function? How do ENM transformations in environmental media and in vivo influence these effects?*** A suite of standard tests for higher-level effects (effects on the community or ecosystem) does not exist. That problem, which is not peculiar to ENMs, presents a serious challenge for modeling ecologic impacts. There is a need for basic research to describe those potential ecosystem impacts with key ENMs, rather than with only high-throughput assays.

Common Issues in Human Health and Ecologic Effects Research

Issues that cut across human and ecologic health include determining the potential mechanisms of toxicity of nanomaterials and how they vary with ENM composition and dose, including developing data so as to correlate in vitro and in vivo responses; understanding effects of chronic exposure to nanomaterials; and obtaining data on multiple end points that precede or do not result in death of cells or organisms.

There are also some common issues regarding experimental design and methods. Access to a library of materials that have a variety of core and surface properties is needed so that a systematic evaluation of ENM properties can be conducted. Sufficient nanomaterials that have different structures and modifications are not available; there is little information about what may be most appropriate to test from an industry or commercial standpoint; and standard negative or positive control materials are not on hand to use for assays and comparisons among laboratories. There is also a need for standardized reference materials.

Dosimetric studies—to understand the transformations of ENMs in vivo—and the ability to characterize ENMs in vivo or in a representative physiologic buffer are needed to correlate the properties of ENMs with their observed effects. For each experiment, information is needed on the key physicochemical properties of ENMs, such as size and size distribution, shape, agglomeration and aggregation state, surface properties (area, charge, reactivity, coating and contaminant chemistry, and defects), solubility (lipid and aqueous), and crystallinity, many of which can change depending on the method of production, during preparation, or storage. Moreover, surface changes will occur when materials are introduced into physiologic media or into an organism. Although data on the impact of such changes on biodistribution and effects are beginning to be generated, a major gap is methods for characterizing the altered surface of nanomaterials after transformation that results from interaction with proteins and lipids at different sites in an organism. Determination of the form of the ENMs that an organism will be exposed to will depend on the fate and transformation of ENMs in the environment (to be assessed with models) and will help to inform effects testing.

Additional challenges arise with the use of dispersants, solvents, or organic carbon, generally accepted and recommended to render nanoparticles either monodispersed or stable in toxicity studies. Such techniques pose a dilemma, in that dispersants will alter surface properties of ENMs, which in turn may alter their interactions with cells and organisms and thus affect the dose that an organism receives or that is delivered to an in vitro cellular assay. Comparisons of results achieved with each dispersion method are needed so that the most appropriate conditions for toxicity testing can be selected.

Toward an Understanding of Systems and Complexity in Nanotechnology-Related Environmental, Health, and Safety Research

The research gaps presented in the prior sections were categorized into questions regarding stages in the source-to-response paradigm of an ENM, but it is critical to recognize the interplay between the questions. That interplay is at the heart of systems science that recognizes that issues in one part of a paradigm can influence outcomes in other parts. This overarching issue has been inadequately addressed in the literature and is best addressed with models (see Chapter 4).

Materials originate at various points along the value chain and life cycle and direct exposures of the environment or organisms to ENMs occur, as described in Chapter 2 (Figure 2-1). The ENMs will have specific surface properties and chemical characteristics (for example, size, shape, chemical composition, and charge). Those properties will determine the types of processes that the materials undergo that can affect their potential for exposure or hazard (for example, attachment to surfaces or dissolution). The present chapter has presented processes that most likely will affect the exposure and hazard potential of ENMs. The impact of those processes on the potential for exposure or hazard can be measured or modeled (see Chapter 4). Models are needed because limited resources do not allow direct measurement of the exposure or toxicity potential of all new ENMs that come to market. However, measurements are needed to construct and validate exposure and toxicity models. Data on all aspects of EHS research regarding ENM properties, processes, and model validation should be collected and stored in a manner that enables data mining and integration with bioinformatic models. The deployment of bioinformatic models will require harmonization of data types and protocols that facilitate sharing. Such models will ultimately enable risk characterization and risk-based decisions to be made on the basis of the properties of ENMs released to the environment. Two examples that illustrate this process follow.

Consider the release of single-walled carbon nanotubes (SWCNT) to the environment. They may be released during manufacture, during use of products containing them, or at the end of their life. Their properties (for example, surface charge, surface functionality, aspect ratio, and presence of adsorbed macromolecules) will affect the fundamental processes that control their exposure potential

TABLE 3-1 Summary of Critical Research Questions

Sources

What types of ENMs are of the highest priority with regards to nano-EHS research?

What are the maximum anticipated amounts of ENMs to which workers, consumers, and ecosystems could be exposed?

How might concentrations of ENMs from different sources apportion themselves in workplace, consumer, and various environmental compartments?

How can ENMs be detected in air, in water, and in complex media, to allow real-time monitoring of ENM sources of exposures?

Modifications and Exposures

Human Health

What conditions will cause ENMs in the gas phase, in liquids, or embedded in solids to become airborne?

What is the ability of certain ENMs (for example, photoactive materials) to persist on skin after application and what are the potential effects?

What types of ENMs can survive the gastrointestinal tract? Do they assimilate intact into the organism?

What are the nature and implications of biomolecular modifications of ENMs?

Ecosystem Health

Under what conditions will ENMs aggregate or disaggregate in relevant environmental media?

How stable are the coatings of ENMs? How does this relate to ENM aggregation and fate?

How rapidly do ENMs dissolve in various relevant environmental media?

What properties of ENMs promote attachment to environmentally relevant surfaces?

How can the fate and transport of ENMs in the environment be fully described and modeled?

Dose, Biodistribution, and Bioavailability

Human Health

What is the most appropriate metric to describe ENM dose?

How does applied dose (for example, dermal, ingestion or inhalation) translate into bioavailability?

Can dosimetric extrapolations between organisms be used to validate in-vitro and in-vivo studies?

What are the most valid quantitative biokinetic models that relate ENM properties to their distribution in organisms, organs, and tissues?

Ecosystem Health

What factors control the distribution of ENMs into biologic compartments in the environment?

How can the environmental dose to an organism be related to bioavailability?

Can ENMs bioaccumulate and to what extent? If so, what properties are most critical for bioaccumulation?

Hazard

Human Health

What biologic effects occur at realistic ENM doses and dose rates? How do ENM properties influence these effects?

How can in-vitro assays be developed and validated so that results are relevant to in-vivo exposures?

How do ENM properties influence toxicologic mechanisms of action?

Ecosystem Health

How can toxicity mechanisms for ENMs be better understood?

How can community and ecosystem level effects be anticipated from single organism tests?

How do properties of ENMs (and their transformations) influence community structure and ecosystem function?

and toxicity. Charge, functional groups, and adsorption of organic macromolecules all increase or decrease attachment of SWCNTs to surfaces, including other SWCNTs, environmental surfaces (such as aquifer media), and cell membranes. The tendency of an ENM to attach to a surface affects aggregation, mobility in porous media, and cellular uptake, which are described in exposure models. Those properties also affect the potential for hazard and are described in hazard models. In that way, the properties of an ENM are correlated with its potential for exposure and toxic effects.

Another example is silver nanoparticles (Ag NPs). Substantial amounts of data are available on the fate, transport, and effects of Ag NPs (Marambio-Jones and Hoek 2010). Ag NPs can undergo oxidative dissolution and dissolve in water and biologic media. Dissolution affects persistence in the environment and exposure potential. Soluble Ag species also affect the toxicity pathways and modes of action of Ag NPs, and it is important to determine whether there is a toxic effect of NPs or the effect is the result of the Ag ion. Thus, dissolution is an important process that affects both exposure and hazard potential of Ag NPs, and it is important to consider factors affecting dissolution rates in different biologic and environmental media (for example, Levard et al. 2011; Liu et al. 2010). The properties of the NPs and media that can be used to predict the rate and extent of dissolution of Ag NPs remain to be fully determined; once they are better understood, the potential exposure to and hazard posed by Ag NPs can be related to them.

REFERENCES

Aggarwal, S.G. 2010. Recent developments in aerosol measurement techniques and the metrological issues. MAPAN 25(3):165-189.

Alexis, F., E. Pridgen, L.K. Molnar, and O.C. Farokhzad. 2008. Factors affecting the clearance and biodistribution of polymeric nanoparticles. Mol. Pharm. 5(4):505-515.

Asbach, C., H. Kaminski, H. Fissan, C. Monz, D. Dahmann, S. Mülhopt, H.R. Paur, H.J. Kiesling, F. Herrmann, M. Voetz, and T.A.J. Kuhlbusch. 2009. Comparison of four mobility particle sizers with different time resolution for stationary exposure measurements. J. Nanopart Res. 11(7):1593-1609.

Asgharian, B., F.J. Miller, and R.P. Subramaniam. 1999. Dosimetry Software to Predict Particle Deposition in Humans and Rats. CIIT Activities 19(3).

Auffan, M., W. Achouak, J. Rose, M.A. Roncato, C. Chaneac, D.T. Waite, A. Masion, J.C. Woicik, M.R. Wiesner, and J.Y. Bottero. 2008. Relation between the redox state of iron-based nanoparticles and their cytotoxicity toward Escherichia coli. Environ. Sci. Technol. 42(17):6730-6735.

Bae, E., H.J. Park, J. Lee, Y. Kim, J. Yoon, K. Park, K. Choi, and J. Yi. 2010. Bacterial cytotoxicity of the silver nanoparticle related to physicochemical metrics and agglomeration properties. Environ. Toxicol. Chem. 29(10):2154-2160.

Balbus, J.M., A.D. Maynard, V.L. Colvin, V. Castranova, G.P. Daston, R.A. Denison, K.L. Dreher, P.L. Goering, A.M. Goldberg, K.M. Kulinowski, N.A. Monteiro-Riviere, G. Oberdörster, G.S. Omenn, K.E. Pinkerton, K.S. Ramos, K.M. Rest,

J.B. Sass, E.K. Silbergeld, and W.A Wong. 2007. Meeting report: Hazard assessment for nanoparticles--report from an interdisciplinary workshop. Environ. Health Perspect. 115(11):1654-1659.

Bazile, D.V., C. Ropert, P. Huve, T. Verrecchia, M. Marlard, A. Frydman, M. Veillard, and G. Spenlehauer. 1992. Body distribution of fully biodegradable [c-14] poly(lactic acid) nanoparticles coated with albumin after parenteral administration to rats. Biomaterials 13(15):1093-1102.

Bello, D., A.J. Hart, K. Ahn, M. Hallock, N. Yamamoto, E.J. Garcia, M.J. Ellenbecker, and B.L. Wardle. 2008. Particle exposure levels during CVD growth and subsequent handling of vertically-aligned carbon nanotube films. Carbon 46(6):974-981.

Bello, D., B.L. Wardle, N. Yamamoto, R.G. deVilloria, E.J. Garcia, A.J. Hart, K. Ahn, M.J. Ellenbecker, and M. Hallock. 2009. Exposure to nanoscale particles and fibers during machining of hybrid advanced composites containing carbon nanotubes. J. Nanopart. Res. 11(1):231-249.

Bernhardt, E.S., B.P. Colman, M.F. Hochella, Jr., B.J. Cardinale, R.M. Nisbet, C.J. Richardson, and L. Yin. 2010. An ecological perspective on nanomaterial impacts in the environment. J. Environ. Qual. 39(6):1954-1965.

Blaser, S.A., M. Scheringer, M. Macleod, and K. Hungerbühler. 2008. Estimation of cumulative aquatic exposure and risk due to silver: Contribution of nano-functionalized plastics and textiles. Sci. Total Environ. 390(2-3):369-409.

Choi, H.S., W. Liu, P. Misra, E. Tanaka, J.P. Zimmer, B.I. Ipe, M.G. Bawendi, and J.V. Frangioni. 2007. Renal clearance of quantum dots. Nat. Biotechnol. 25(10):1165-1170.

Cui, D., F. Tian, C.S. Ozkan, M. Wang, and H. Gao. 2005. Effect of single wall carbon nanotubes on human HEK293 cells. Toxicol. Lett. 155(1):73-85.

Denison, R. 2005. Statement of Richard D. Denison, Senior Scientist, Environmental Defense, before the House of Representatives Committee on Science Concerning Environmental and Safety Impacts of Nanotechnology: What Research is Needed? November 17, 2005 [online]. Available: http://www.edf.org/documents/5136_DenisonHousetestimonyOnNano.pdf [accessed May 24, 2011].

Drinker, P, R.M. Thomson, and J.L. Finn. 1927. Metal fume fever. II. Resistance acquired by inhalation of zinc oxide on two successive days. J. Ind. Hyg. 9(3):98-105.

EFSA (European Food Safety Authority). 2009. Scientific Opinion: The Potential Risks Arising from Nanoscience and Nanotechnologies on Food and Feed Safety. EFSA J. 958:1-39 [online]. Available: http://www.efsa.europa.eu/fr/scdocs/doc/sc_op_ej958_nano_en,0.pdf [accessed Apr. 14, 2011].

EPA (U.S. Environmental Protection Agency). 1998. Guidelines for Ecological Risk Assessment. EPA/630/R-95/002F. Risk Assessment Forum, U.S. Environmental Protection Agency, Washington, DC. April 1998 [online]. Available: http://www.epa.gov/raf/publications/pdfs/ECOTXTBX.PDF [accessed Apr. 20, 2011].

EPA (U.S. Environmental Protection Agency). 2002. Proceedings EPA Nanotechnology and the Environment: Applications and Implications STAR Progress Review Workshop, August 28-29, 2002, Arlington, VA. Office of Research and Development, National Center for Environmental Research, U.S. Environmental Protection Agency [online]. Available: http://www.epa.gov/ncer/publications/workshop/nano_proceed.pdf [accessed May 2, 2011].

FAO/WHO (Food and Agriculture Organization of the United Nation and World Health Organization). 2009. FAO/WHO Expert Meeting on the Application of Nanotechnologies in the Food and Agriculture Sectors: Potential Food Safety Implications.

Meeting Report. Food and Agriculture Organization of the United Nation, Rome, Italy, and World Health Organization, Geneva, Switzerland [online]. Available: http://www.fao.org/ag/agn/agns/files/FAO_WHO_Nano_Expert_Meeting_Report_Final.pdf [accessed Oct. 12, 2011].

Fortner, J.D., D.Y. Lyon, C.M. Sayes, A.M. Boyd, J.C. Falkner, E.M. Hotze, L.B. Alemany, Y.J. Tao, W. Guo, K.D. Ausman, V.L. Colvin, and J.B. Hughes. 2005. C_{60} in water: Nanocrystal formation and microbial response. Environ Sci. Technol. 39(11):4307-4316.

Gopee, N.V., D.W. Roberts, P. Webb, C.R. Cozart, P.H. Siitonen, A.R. Warbritton, W.W. Yu, V.L. Colvin, N.J. Walker, and P.C. Howard. 2007. Migration of intradermally injected quantum dots to sentinel organs in mice. Toxicol. Sci. 98(1):249-257.

Gottschalk, F., T. Sonderer, R.W. Scholz, and B. Nowack. 2009. Modeled environmental concentrations of engineered nanomaterials (TiO_2, ZnO, Ag, CNT, fullerenes) for different regions. Environ. Sci. Technol. 43(24):9216-9222.

Gottschalk, F., R.W. Scholz, and B. Nowack. 2010. Probabilistic material flow modeling for assessing the environmental exposure to compounds: Methodology and an application to engineered nano-TiO_2 particles. Environ. Modell. Softw. 25(3):320-332.

Guo, J., X. Zhang, Q. Li, and W. Li. 2007. Biodistribution of functionalized multiwall carbon nanotubes in mice. Nucl. Med. Biol. 34(5):579-583.

Han, J.H., E.J. Lee, J.H. Lee, K.P. So, Y.H. Lee, G.N. Bae, S.B. Lee, J.H. Ji, M.H. Cho, and I.J. Yu. 2008. Monitoring multiwalled carbon nanotube exposure in carbon nanotube research facility. Inhal. Toxicol. 20(8):741-749.

Heckmann, L.H., J. Baas, and T. Jager. 2010. Time is of the essence. Environ. Toxicol. Chem. 29(6):1396-1398.

Hirose, A., T. Nishimura, and J. Kanno. 2009. Research strategy for evaluation methods of the manufactured nanomaterials in NIHS and importance of the chronic health effects studies [in Japanese]. Kokuritsu Iyakuhin Shokuhin Eisei Kenkyusho Hokoku 127:15-27.

Hotze, E.M., T. Phenrat, and G.V. Lowry. 2010. Nanoparticle aggregation: Challenges to understanding transport and reactivity in the environment. J. Environ. Qual. 39(6):1909-1924.

Hou, W.C., L. Kong, K. Wepasnick, R.G. Zepp, D.H. Fairbrother, and C.T. Jafvert. 2010. Photochemistry of aqueous C_{60} clusters: Wavelength dependency and product characterization. Environ. Sci. Technol. 44(21):8121-8127.

ICON (International Council on Nanotechnology). 2011. Nano-EHS Database Analysis Tool [online]. Available: http://cohesion.rice.edu/centersandinst/icon/report.cfm [accessed May 27, 2011].

ICRP (International Commission on Radiological Protection). 1994. Human Respiratory Tract Model for Radiological Protection. ICRP Publication 66. Ann. ICRP 24(1-3).

ISO (International Organization for Standardization). 2008. Nanotechnologies. Terminology and Definitions for Nano-Objects. Nanoparticle, Nanofibre and Nanoplate. ISO/TS 27687:2008. Geneva: ISO.

Jager, T., T. Vandenbrouck, J. Baas, W.M. De Coen, and S.A. Kooijman. 2010. A biology-based approach for mixture toxicity of multiple endpoints over the life cycle. Ecotoxicology 19(2):351-361.

Jarvie, H.P., H. Al-Obaidi, S.M. King, M.J. Bowes, M.J. Lawrence, A.F. Drake, M.A. Green, and P.J. Dobson. 2009. Fate of silica nanoparticles in simulated primary wastewater treatment. Environ. Sci. Technol. 43(22):8622-8628.

Jeong, C.H., and G.J. Evans. 2009. Inter-comparison of a Fast Mobility Particle Sizer and a Scanning Mobility Particle Sizer incorporating an ultrafine water-based condensation particle counter. Aerosol Sci. Technol. 43(4):364-373.

Johnston, H.J., G. Hutchison, F.M. Christensen, S. Peters, S. Hankin, and V. Stone. 2010. A review of the in vivo and in vitro toxicity of silver and gold particulates: Particle attributes and biological mechanisms responsible for the observed toxicity. Crit. Rev. Toxicol. 40(4):328-346.

Johnston, J.M., M. Lowry, S. Beaulieu, and E. Bowles. 2010. State-of-the-Science Report on Predictive Models and Modeling Approaches for Characterizing and Evaluating Exposure to Nanomaterials. EPA/600/R-10/129. U.S. Environmental Protection Agency, Washington, DC [online]. Available: http://www.epa.gov/athens/publications/reports/Johnston_EPA600R10129_State_of_Science_Predictive_Models.pdf [accessed Nov. 23, 2011].

Kim, B., C.S. Park, M. Murayama, and M.F. Hochella. 2010. Discovery and characterization of silver sulfide nanoparticles in final sewage sludge products. Environ. Sci. Technol. 44(19):7509-7514.

Kiser, M.A., P. Westerhoff, T. Benn, Y. Wang, J. Perez-Rivera, and K. Hristovski. 2009. Titanium removal and release from wastewater treatment plants. Environ. Sci. Technol. 43(17):6757-6763.

Klaper, R., J. Crago, J. Barr, D. Arndt, K. Setyowati, and J. Chen. 2009. Toxicity biomarker expression in daphnids exposed to manufactured nanoparticles: Changes in toxicity with functionalization. Environ. Pollut. 157(4):1152-1156.

Kolosnjaj-Tabi, J., K.B. Hartman, S. Boudjemaa, J.S. Ananta, G. Morgant, H. Szwarc, L.J. Wilson, and F. Moussa. 2010. In vivo behavior of large doses of ultrashort and full-length single-walled carbon nanotubes after oral and intraperitoneal administration to Swiss mice. ACS Nano 4(3):1481-1492.

Kreyling, W.G., M. Semmler, F. Erbe, P. Mayer, S. Takenaka, H. Schulz, G. Oberdörster, and A. Ziesenis. 2002. Translocation of ultrafine insoluble iridium particles from lung epithelium to extrapulmonary organs is size dependent but very low. J. Toxicol. Environ. Health A. 65(20):1513-1530.

Kreyling, W.G., M. Semmler-Behnke, J. Seitz, W. Scymczak, A. Wenk, P. Mayer, S. Takenaka, and G. Oberdörster. 2009. Size dependence of the translocation of inhaled iridium and carbon nanoparticle aggregates from the lung of rats to the blood and secondary target organs. Inhal Toxicol. 21(suppl. 1):55-60.

Kuhlbusch, T.A., S. Neumann, and H. Fissan. 2004. Number size distribution, mass concentration, and particle composition of PM_1, $PM_{2.5}$, and PM_{10} in bag filling areas of carbon black production. J. Occup. Environ. Hyg. 1(10):660-671.

Lacerda, L., M. A. Herrero, K. Venner, A. Bianco, M. Prato, and K. Kostarelos. 2008a. Carbon-nanotube shape and individualization critical for renal excretion. Small 4(8):1130-1132.

Lacerda, L., A. Soundararajan, R. Singh, G. Pastorin, K.T. Al-Jamal, J. Turton, P. Frederik, M.A. Hererro, S. Li, A. Bao, D. Emfietzoglou, S. Mather, W.T. Phillips, M. Prato, A. Bianco, B. Goins, and K. Kostarelos. 2008b. Dynamic imaging of functionalized multi-walled carbon nanotube systemic circulation and urinary excretion. Adv. Mater. 20(2):225-230.

Lee, J.H., S.B. Lee, G.N. Bae, K.S. Jeon, J.U. Yoon, J.H. Ji, J.H. Sung, B.G. Lee, J.H. Lee, J.S. Yang, H.Y. Kim, C.S. Kang, and I.J. Yu. 2010. Exposure assessment of carbon nanotube manufacturing workplaces. Inhal. Toxicol. 22(5):369-381.

Levard, C., B.C. Reinsch, F.M. Michel, C. Oumahi, G.V. Lowry, and G.E. Brown. 2011. Sulfidation processes of PVP-coated silver nanoparticles in aqueous solution: Impact on dissolution rate. Environ. Sci. Technol. 45(12):5260-5266.

Lewinski, N., V. Colvin, and R. Drezek. 2008. Cytotoxicity of nanoparticles. Small 4(1):26-49.

Li, Z., K. Greden, P.J. Alvarez, K.B. Gregory, and G.V. Lowry. 2010. Adsorbed polymer and NOM limits adhesion and toxicity of nano scale zerovalent iron to E. coli. Environ. Sci. Technol. 44(9):3462-3467.

Liu, J., D.A. Sonshine, S. Shervani, and R.H. Hurt. 2010. Controlled release of biologically active silver from nanosilver surfaces. ACS Nano 4(11):6903-6913.

Lux Research. 2009. The Recession's Ripple Effect on Nanotech. Lux Research, June 9, 2009 [online]. Available: https://portal.luxresearchinc.com/reporting/research/document_excerpt/4995 [accessed Dec. 5, 2011].

Lynch, I., T. Cedervall, M. Lundqvist, C. Cabaleiro-Lago, S. Linse, and K.A. Dawson. 2007. The nanoparticle-protein complex as a biological entity: A complex fluids and surface science challenge for the 21st century. Adv. Colloid Interface Sci. (134-135):167-174.

Lyon, D.Y., and P.J. Alvarez. 2008. Fullerene water suspension (nC_{60}) exerts antibacterial effects via ROS-independent protein oxidation. Environ. Sci. Technol. 42(21):8127-8132.

Lyon, D.Y., D.A. Brown, and P.J. Alvarez. 2008. Implications and potential applications of bactericidal fullerene water suspensions: Effect of nC_{60} concentration, exposure conditions and shelf life. Water Sci. Technol. 57(10):1533-1538.

Maldiney, T., C. Richard, J. Seguin, N. Wattier, M. Bessodes, and D. Scherman. 2011. Effect of core diameter, surface coating, and PEG chain length on the biodistribution of persistent luminescence nanoparticles in mice. ACS Nano. 5(2):854-862.

Mahendra, S. H. Zhu, V.L. Colvin, and P.J. Alvarez. 2008. Quantum dot weathering results in microbial toxicity. Environ. Sci. Technol. 42(24):9424-9430.

Marambio-Jones, C., and E.M.V. Hoek . 2010. A review of the antibacterial effects of silver nanomaterials and potential implications for human health and the environment. J. Nanopart. Res. 12(5):1531-1551.

Maynard, A.D., P.A. Baron, M. Foley, A.A. Shvedova, E.R. Kisin, and V. Castranova. 2004. Exposure to carbon nanotube material: Aerosol release during the handling of unrefined single-walled carbon nanotube material. J. Toxicol. Environ. Health A 67(1):87-107.

McMurry, P.H. 2000. A review of atmospheric aerosol measurements. Atmos Environ. 34(12-14):1959-1999.

Methner, M., L. Hodson, A. Dames, and C. Geraci. 2010. Nanoparticle Emission Assessment Technique (NEAT) for the identification and measurement of potential inhalation exposure to engineered nanomaterials, Part B: Results from 12 field studies. J. Occup. Environ. Hyg. 7(3):163-176.

Moghimi, S.M., and M. Moghimi. 2008. Enhanced lymph node retention of subcutaneously injected IgG1-PEG2000-liposomes through pentameric IgM antibody-mediated vesicular aggregation. Biochim. Biophys. Acta 1778(1):51-55.

Monopoli, M.P., D. Walczyk, A. Campbell, G. Elia, I. Lynch, F.B. Bombelli, and K.A. Dawson. 2011. Physical-chemical aspects of protein corona: Relevance to in vitro and in vivo biological impacts of nanoparticles. J. Am. Chem. Soc. 133(8):2525-2534.

Mortensen, L.J., G. Oberdörster, A.P. Pentland, L.A. DeLouise. 2008. In vivo skin penetration of quantum dot nanoparticles in the murine model: The effect of UVR. Nano Lett. 8(9):2779-2787.
Mueller, N.C., and B. Nowack. 2008. Exposure modeling of engineered nanoparticles in the environment. Environ. Sci. Technol. 42(12):4447-4453.
Müller, R.H. and C.M. Keck. 2004. Drug delivery to the brain-realization by novel drug carriers. J. Nanosci. & Nanotechnol. 4(5):471-483.
Mylon, S.E., K.L. Chen, and M. Elimelech. 2004. Influence of natural organic matter and ionic composition on the kinetics and structure of hematite colloid aggregation: Implications to iron depletion in estuaries. Langmuir 20(21):9000-9006.
Nel, A., T. Xia, L. Madler, and N. Li. 2006. Toxic potential of materials at the nanolevel. Science 311(5761):622-627.
NNI (National Nanotechnology Initiative). 2011a. NNI Meetings and Workshops [online]. Available: http://www.nano.gov/events/meetings-workshops [accessed May 4, 2011].
NNI (National Nanotechnology Initiative). 2011b. NNI Centers and Networks [online]. Available: http://www.nano.gov/initiatives/government/research-centers [accessed May 4, 2011].
NRC (National Research Council). 2009. Science and Decisions: Advancing Risk Assessment. Washington DC: The National Academies Press.
Oberdörster, G. 1989. Dosimetric principles for extrapolating results of rat inhalation studies to humans, using an inhaled Ni compound as an example. Health Physics 57(suppl 1):213-220.
Oberdörster, G., Z. Sharp, V. Atudorei, A. Elder, R. Gelein, W. Kreyling, and C. Cox. 2004. Translocation of inhaled ultrafine particles to the brain. Inhal. Toxicol.. 16(6-7):437-445.
Oberdörster, G., V. Stone, and K. Donaldson. 2007. Toxicology of nanoparticles: A historical perspective. Nanotoxicology 1(1):2-25.
Oberdörster, G. 2011. Correlating in vitro and in vivo nanotoxicity: limitations and challenges for risk assessment. Presented at the Society of Toxicology meeting. March 6-10, 2011. Washington, DC.
Ohl, L., M. Mohaupt, N. Czeloth, G. Hintzen, Z. Kiafard, J. Zwirner, T. Blankenstein, G. Henning, and R. Förster. 2004. CCR7 governs skin dendritic cell migration under inflammatory and steady-state conditions. Immunity 21(2):279-288.
Pauluhn, J. 2010. Multi-walled carbon nanotubes (Baytubes): Approach for derivation of occupational exposure limit. Regul. Toxicol. Pharmacol. 57(1):78-89.
PEN (Project on Emerging Nanotechnologies). 2011. Consumer Products: An Inventory of Consumer Products Currently on the Market. Project on Emerging Nanotechnologies [online]. Available: http://www.nanotechproject.org/inventories/consumer/ [accessed May 3, 2011].
Petosa, A.R., D.P. Jaisi, I.R. Quevedo, M. Elimelech, and N. Tufenkji. 2010. Aggregation and deposition of engineered nanomaterials in aquatic environments: Role of physicochemical interactions. Environ. Sci. Technol. 44(17):6532-6549.
Phenrat, T., H.J. Kim, F. Fagerlund, T. Illangasekare, R.D. Tilton, and G.V. Lowry. 2009a. Particle size distribution, concentration, and magnetic attraction affect transport of polymer-modified Fe^0 nanoparticles in sand columns. Environ. Sci. Technol. 43(13):5079-5085.
Phenrat, T., T.C. Long, G.V. Lowry, and B. Veronesi. 2009b. Partial oxidation ("aging") and surface modification decrease the toxicity of nanosized zerovalent iron. Environ. Sci. Technol. 43(1):195-200.

Phenrat, T., J.E. Song, C.M. Cisneros, D.P. Schoenfelder, R.D. Tilton, and G.V. Lowry. 2010. Estimating attachment of nano- and submicrometer-particles coated with organic macromolecules in porous media: Development of an empirical model. Environ. Sci. Technol. 44(12):4531-4538.

Prow, T.W., J.E. Grice, L.L. Lin, R. Faye, M. Butler, W. Becker, E.M.T. Wurm, C. Yoong, T.A. Robertson, H.P. Soyer, and M.S. Roberts. 2011. Nanoparticles and microparticles for skin drug delivery. Advan. Drug Deliv. Rev. 63(6):470-491.

Robichaud, C.O., A.E. Uyar, M.R. Darby, L.G. Zucker, and M.R. Wiesner. 2009. Estimates of upper bounds and trends in nano-TiO_2 production as a basis for exposure assessment. Environ. Sci. Technol. 43(12):4227-4233.

Rushton, E.K., J. Jiang, S.S. Leonard, S. Eberly, V. Castranova, P. Biswas, A. Elder, X. Han, R. Gelein, J. Finkelstein, and G. Oberdörster. 2010. Concept of assessing nanoparticle hazards considering nanoparticle dosimetric and chemical/biological response metrics. J. Toxicol. Environ. Health A 73(5):445-461.

Sahu, M., and P. Biswas. 2010. Size distributions of aerosols in an indoor environment with engineered nanoparticle synthesis reactors operating under different scenarios. J. Nano. Res. 12(3):1055-1964.

Saleh, N.B., L.D. Pfefferle, and M. Elimelech. 2008. Aggregation kinetics of multiwalled carbon nanotubes in aquatic systems: Measurements and environmental implications. Environ. Sci. Technol. 42(21):7963-7969.

Saleh, N.B., L.D. Pfefferle, and M. Elimelech. 2010. Influence of biomacromolecules and humic acid on the aggregation kinetics of single-walled carbon nanotubes. Environ. Sci. Technol. 44(7):2412-2418.

Salonen, E., S. Lin, M.L. Reid, M. Allegood, X. Wang, A.M. Rao, I. Vattulainen, and P.C. Ke. 2008. Real-time translocation of fullerene reveals cell contraction. Small 4(11):1986-1992.

Schluep, T., J. Hwang, I.J. Hildebrandt, J. Czernin, C.H. Choi, C.A. Alabi, B.C. Mack, and M.E. Davis. 2009. Pharmacokinetics and tumor dynamics of the nanoparticle IT-101 from PET imaging and tumor histological measurements. Proc. Natl. Acad. Sci. USA. 106(27):11394-11399.

Schneider, M., F. Stracke, S. Hansen, and U.F. Schaefer. 2009. Nanoparticles and their interactions with the dermal barrier. Dermato-Endocrinology 1(4):197-206.

Semmler-Behnke, M., S. Takenaka, S. Fertsch, A. Wenk, J. Seitz, P. Mayer, G. Oberdörster, and W.G. Kreyling. 2007. Efficient elimination of inhaled nanoparticles from the alveolar region: Evidence for interstitial uptake and subsequent reentrainment onto airways epithelium. Environ. Health Perspect. 115(5):728-733.

Semmler-Behnke, M., W.G. Kreyling, J. Lipka, S. Fertsch, A. Wenk, S. Takenaka, G. Schmid, and W. Brandau. 2008. Biodistribution of 1,4-and 18-nm gold particles in rats. Small 4(12):2108-2111.

Slikker, W., Jr., M.E. Andersen, M.S. Bogdanffy, J.S. Bus, S.D. Cohen, R.B. Conolly, R.M. David, N.G. Doerrer, D.C. Dorman, D.W. Gaylor, D. Hattis, J.M. Rogers, W. Setzer, J.A. Swenberg, and K. Wallace. 2004a. Dose-dependent transitions in mechanisms of toxicity: Case studies. Toxicol. Appl. Pharmacol. 201(3):226-294.

Slikker, W., Jr., M.E. Andersen, M.S. Bogdanffy, J.S. Bus, S.D. Cohen, R.B. Conolly, R.M. David, N.G. Doerrer, D.C. Dorman, D.W. Gaylor, D. Hattis, J.M. Rogers, R. Woodrow Setzer, J.A. Swenberg, and K. Wallace. 2004b. Dose-dependent transitions in mechanisms of toxicity. Toxicol. Appl. Pharmacol. 201(3):203-225.

Smijs, T.G., and J.A. Bouwstra. 2010. Focus on skin as a possible port of entry for solid nanoparticles and the toxicological impact. J. Biomed. Nanotechnol. 6(5):469-484.

Tang, J., L. Xiong, S. Wang, J. Wang, L. Liu, J. Li, F. Yuan, and T. Xi. 2009. Distribution, translocation and accumulation of silver nanoparticles in rats. J. Nanosci. Nanotechnol. 9(8):4924-4932.

Tong, Z., M. Bischoff, L. Nies, B. Applegate, and R.F. Turco. 2007. Impact of fullerene C_{60} on a soil microbial community. Environ. Sci. Technol. 41(8):2985-2991.

Tsai, S.J., M. Hofmann, M. Hallock, E. Ada, J. Kong, and M. Ellenbecker. 2009. Characterization and evaluation of nanoparticle release during the synthesis of single-walled and multi-walled carbon nanotubes by chemical vapor deposition. Environ. Sci. Technol. 43(15):6017-6023.

Walczyk, D., F.B. Bombelli, M.P. Monopoli, I. Lynch, and K.A. Dawson. 2010. What the cell "sees" in bionanoscience. J. Am. Chem. Soc. 132(16):5761-5768.

Wang, C.Y., W.E. Fu, H.L. Lin, and G.S. Peng. 2007. Preliminary study on nanoparticle sizes under the APEC technology cooperative framework. Meas. Sci. Technol. 18(2):487-495.

Wiesner, M.R., G.V. Lowry, K.L. Jones, M.F. Hochella, R.T. Di Giulio, E. Casman, and E.S. Bernhardt. 2009. Decreasing uncertainties in assessing environmental exposure, risk, and ecological implications of nanomaterials. Environ. Sci. Technol. 43(17):6458-6462.

Wittmaack, K. 2007. In search of the most relevant parameter for quantifying lung inflammatory response to nanoparticle exposure: Particle number, surface area, or what? Environ. Health Perspect. 115(2):187-194.

Xia, T., M. Kovochich, J. Brant, M. Hotze, J. Sempf, T. Oberley, C. Sioutas, J.I. Yeh, M.R. Wiesner, and A.E. Nel. 2006. Comparison of the abilities of ambient and manufactured nanoparticles to induce cellular toxicity according to an oxidative stress paradigm. Nano Lett. 6(8):1794-1807.

Zeliadt, N. 2010. Silver beware: Antimicrobial nanoparticles in soil may harm plant life. Scientific American, August 9, 2010 [online]. Available: http://www.scientificamerican.com/article.cfm?id=silver-beware-antimicrobial-nanoparticles-in-soil-may-harm-plant-life [accessed July 26, 2011].

4

New Tools and Approaches for Identifying Properties of Engineered Nanomaterials That Indicate Risks

This chapter articulates needs for tool development for exploring how properties of engineered nanomaterials (ENMs) influence critical biologic and environmental interactions (see Figure 2-1). The research needs are directed at the gaps in evidence presented in Chapter 3 and are based on the conceptual framework for assessing risks described in Chapter 2. The primary needs are access to nanomaterials for hypothesis-testing and for assessing exposure to and effects of ENMs; methods for characterizing materials, including methods for detecting, quantifying, and characterizing ENMs in environmental and biologic samples; exposure and toxicity-testing methods and reporting standards; exposure and effects modeling; and informatics for developing comprehensive predictive models of exposure, hazards, and risk. Informatics is defined here as the infrastructure and information science and technology needed to integrate data, information, and knowledge on the environmental, health, and safety (EHS) aspects of nanotechnology. An overall purpose of informatics in this context is to organize data so that they can be mined to determine how nanomaterial properties affect their exposure and hazard potential and to estimate overall risks to the environment and human health. (The research needs presented here are summarized according to categories of tools at the end of this chapter, Table 4-1.)

CHARACTERIZED NANOMATERIALS FOR NANOTECHNOLOGY-RELATED ENVIRONMENTAL, HEALTH, AND SAFETY RESEARCH

Identifying ENM properties that influence biologic and environmental interactions will require well-characterized libraries of materials for hypothesis-testing and reference or standard test materials that may be used as benchmarks for comparison among studies, to validate protocols or measurements, or to test specific hypotheses related to material properties and specific outcomes (for example, mobility in the environment or toxic responses). The lack of widespread access to such materials and the lack of agreement as to which materials to consider as standards slows progress toward linking properties of ENMs with their effects, makes comparisons among studies difficult, and limits the utility of data collected for informatics (see section "Barriers to Informatics").

To characterize correlations between nanomaterial properties and the key interactions or end points in humans and the environment, several tools are needed, including adequately characterized materials that have different properties, appropriate assays for examining interactions or end points, and experimental data of sufficient breadth and depth for assessing correlations between nanomaterial properties and the behavior of the materials. Materials needed for developing those correlations are in four general categories, which are described below. Each type must be characterized sufficiently for test results to be reproducible and for correlations between observed effects and material structure and composition to be established and ultimately used to predict effects of new materials on the basis of knowledge of their structure and composition.

Research or Commercial Samples

These samples may be available from R&D teams or from materials that are near commercialization or in commerce. Many EHS studies have been conducted with such materials because of their availability and because people or the environment may be exposed to these materials. The material definition and characterization metrics needed for nanomaterial research and commercial use are typically different from those needed to study material-effect correlations, and the former materials often do not have the definition, purity, or characterization needed for research purposes. It is important to study the biologic and ecologic effects of the commercial materials, as such materials (and their impurities) have the greatest potential compared to other types of materials to be released into the environment (Alvarez et al. 2009; Gottschalk and Nowack 2011). However, there are limitations to the use of commercial materials in the development of predictive models. The materials are generally insufficiently characterized; when they are studied in isolation, the polydispersity and lot-to-lot variation in their properties make them unsuitable for developing data that can be used for prediction. For greater utility in prediction, material characterization that is specific to EHS research should be conducted in addition to that carried out by material researchers or producers (Bouwmeester et al. 2011).

Reference Materials

Reference materials are developed for hypothesis-driven research or for use as benchmarks to compare results among various tests or assays or among laboratories. They are designed and characterized so that material characteristics can be linked to biologic-nanotechnologic or ecologic-nanotechnologic interactions or end points. Reference materials are often highly purified to reduce or eliminate the effects of impurities on responses (Oostingh et al. 2011). They may not attain the same level of scrutiny as standards (see discussion below), but they require a smaller investment of time and resources to develop. Sources of these materials include academic and government research laboratories (National Institute of Standards and Technology), commercial suppliers (for example, nanoComposix, Nanoprobes, Inc., and Strem Chemicals, Inc.), and international harmonization efforts (such as the Organisation for Economic Co-operation and Development and the International Alliance for NanoEHS Harmonization). Standard or reference materials can be used to compare test or measurement results among laboratories or to compare the results from different tests or measurements. However, because these materials typically represent specific, narrow structural types that are not easily manipulated to access a broad range of structural features, it is difficult to develop more general design rules from studies of these materials.

Libraries

Libraries are collections of reference materials in which structural or compositional variables are systematically varied throughout a series of members of the library. For example, the nanoparticle core material and size might be kept constant while a surface coating varies in its external charge—positively, negatively, or not at all. Libraries allow the influence of nanomaterial structure and composition on biologic or ecologic effects to be explored so that quantitative structure-activity relationships can be determined. Libraries also facilitate exploration of hypotheses related to material-effect correlations. To serve that purpose, libraries should be appropriately defined and characterized as described above for reference materials. Ideally, the materials in libraries have sufficient range and granularity across the structural or compositional measures of interest. Given the importance of detailed characterization for establishing cause-effect correlations, characterization data on each sample lot need to be provided with each sample.

Standards

Standards are samples that have been thoroughly tested to support laboratory comparisons or to calibrate and harmonize measurements conducted in different laboratories. They typically are prepared and provided for by standard-setting organizations or agencies (for example, the National Institute of Standards and Technology). The benefits of developing standard materials that meet the criteria for definition and characterization are clear; however, the time (years) and expense of developing such standards sometimes restrict their use in EHS studies.

Research Needs for Providing Well-Characterized Nanomaterials for Nanotechnology-Related Environmental, Health, and Safety Research

- *Development of characterized, reproducible, but not necessarily uniform, "real-world" materials for testing.*
- *Development of libraries of uniform, well-characterized reference materials of varied size, shape, aspect ratio, surface charge, and surface functionality.*
- *Development of standard materials for calibrating various assays and measurement tools.*
- *Development of new synthetic methods and postsynthesis separation and purification methods for accessing the different types of materials, reducing polydispersity, and decreasing lot-to-lot variability and for efficiently removing undesirable impurities from nanomaterials without causing their decomposition or agglomeration.*

TOOLS, STANDARDIZED CHARACTERIZATION METHODS, AND NOMENCLATURE OF ENGINEERED NANOMATERIALS

Protocols for Measuring and Reporting a Minimum Set of Material Properties for Pristine Engineered Nanomaterials Used in Nanotechnology-Related Environmental, Health, and Safety Research

With regard to characterization of research and commercial samples for EHS testing, there is a need for systematic approaches for adequately and systematically defining the structure, composition (including surface chemistry), and purity of samples so that data reported through the nanotechnology-related EHS research community ultimately can be used to correlate structure and composition of nanomaterials with their behaviors and effects. Most of the tools needed to accomplish that goal are available for pristine[1] starting materials (Hassellöv et al. 2008). One exception is the lack of tools for characterizing the details of the surface chemistry of nanoparticles, including defects in surface layers, mixtures of bound molecules, and conformation of the adsorbed layer of organic macromolecules of high molecular weight. That type of characterization should form the basis of a working definition (or nomenclature) for the material. For example, the intent would be to move from labeling a material as "gold nanoparticles" to the more specific designation of "mercaptopropionic acid stabilized 1.5 ± 0.4 nm gold nanoparticles." Each material lot needs to be characterized in that way (because of variations from batch to batch). Polydisperse and impure samples (for example, materials that have varied chemical composition

---

[1]Pristine refers to the nanomaterial as manufactured, before any alterations in the environment.

or that contain endotoxins) are inherently more complicated to characterize because they are mixtures. For commercial or research samples, a material should be characterized to assess purity and size distribution in the state in which it is provided to researchers. Reference materials and libraries may require extensive purification to remove impurities or to decrease polydispersity that complicates data interpretation and characterization.

Despite concerted efforts to establish a minimum set of standard properties to define ENMs, there is still lack of agreement in the research community as to what constitutes this minimum set of properties. Yet there has been some progress in demonstrating that there is overlap in their nanomaterial properties (MINChar Initiative 2009; Boverhof and David 2010). Without agreement on the properties and how they can be communicated, with full participation of the nanomaterial-EHS research community, it will not be possible to define the starting materials for nanomaterial-EHS research adequately or to create "classes" of ENMs that have similar surface chemistries and behaviors. Therefore efforts to compare results among studies with informatics or other approaches will be hindered (see section "Barriers to Informatics"). Ultimately, a classification of ENMs will probably be needed for regulatory purposes, but the criteria for what constitutes a "class" have not been determined.

Because of the complexity of nanomaterial structures and compositions, a wide array of techniques is typically needed to characterize each new nanomaterial adequately. Each technique provides a partial definition of the material. For example, for a ligand-stabilized inorganic nanoparticle, transmission electron microscopy (TEM) and small-angle x-ray scattering can be used to define nanoparticle cores (von der Kammer et al. 2012); x-ray photoelectron spectroscopy and Fourier-transform infrared (FTIR) spectroscopy define surface chemistry; atomic-force microscopy provides information about the overall dimension of the core plus shell; thermal gravimetric analysis provides the ratio of ligand mass to core mass; and solution methods, such as nuclear magnetic resonance spectrometry, can be used to detect small-molecule impurities. Because such exhaustive characterization of each nanomaterial sample is expensive and time-consuming, minimal characterization sets have been proposed (for example, Boverhof and David 2010). One approach is to make the same comprehensive or subset of measurements for every material; however, this approach can lead to unneeded measurements of some materials or insufficient characterization of others. Other approaches seek to determine the minimum material properties that need to be defined to describe materials used in nanotechnology-related EHS studies adequately and should address at least physical dimensions, composition (including surface chemistry), and purity (MINChar Initiative 2009; Richman and Hutchison 2009). From these approaches key material descriptors should emerge that will facilitate attribution of material effects, data-sharing, and comparison of properties and effects between samples.

In addition to assessment of pristine material samples and dry powders, analytic methods should include characterization of ENMs in various reference suspension media that reflect real-world fluid suspension media and concentrations (for example, water, phosphate-buffered solution, lung fluid, and plasma) because ENM properties are determined in part by the dispersing fluid and ENM concentration (Oberdörster et al. 2005). Reactivity measurements are also needed and could include redox activity and reactive-oxygen species generation.

Protocols and methods will need to be specific to a nanomaterial's characteristics, including particle type, size, shape, coating type, and media type, because not all methods will be applicable to all types of ENMs. There are some key issues that if left unaddressed lead to problems, including methods for dispersing nanoparticles in media, protocols for reproducibly preparing samples for analysis and investigation, and approaches to using multiple instruments to cross-check and confirm results from techniques that may provide only partial answers. There is a need for widely accepted protocols for sample preparation and measurements; for example, see the National Cancer Institute Nanotechnology Characterization Laboratory's effort to develop and publish assay cascade protocols (including NIST/NCL 2010). The sensitivity of the protocols to the array of variables that may affect their outcome (for example, solution pH and energy input for creating a dispersion) should be determined and reported as part of the protocols.

Tools and methods are needed to characterize the surface properties of ENMs better in situ or in vivo. As discussed in Chapter 3, these properties will depend on the media in which they are dispersed so methods should be tailored to the exposure conditions. The surface properties of ENMs will determine their interactions with environmental and biologic media. Many tools are available to characterize size, elemental composition, and structure, but fewer are capable of characterizing only the surfaces of ENMs. Surface curvature, roughness, crystal faces, and defects may all affect the physical, chemical, and toxicologic properties of an ENM; it is not possible to characterize those features adequately with existing microscopic and spectroscopic techniques (for example, electron spectroscopy for chemical analysis, TEM, and FTIR). Surface functional groups—such as adsorbed or grafted surfactants, polymers, polyelectrolytes, proteins, and natural organic matter (NOM)—can prevent or enhance agglomeration and deposition (Phenrat et al. 2008; Saleh et al. 2008; Jarvie et al. 2009), toxicity (Gao et al. 2005; Nel et al. 2009; Phenrat et al. 2009), and bioavailability (Kreuter 1991). Despite the influence of bound coatings on ENM behavior, methods for readily measuring the distribution and, more important, the conformation of the bound species on the surface of ENMs are not widely available. Cryoelectron microscopy combined with computational methods can provide information on conformation of antibodies or other molecules, but these methods are time-consuming, and results can be influenced by sample-preparation methods. Methods for measuring those features in vivo, in vitro, or in situ do not exist and their development is necessary to begin to correlate the in situ properties of ENMs with their behavior and effects.

Research Needs for Developing Protocols for Measuring and Reporting a Minimum Set of Material Properties for Pristine Engineered Nanomaterials Used in Nanotechnology-Related Environmental, Health, and Safety Research

- *Identify agreed-on minimum characterization principles to develop standardized descriptors for ENMs related to the key physical characteristics of the materials that can be used to describe materials for data-reporting and informatics and for cross-referencing nomenclatures (that is, nanomaterial vocabularies and ontologies).*
- *Determine best practices for characterizing groups of particle types (for example, by chemical composition or chemical-surface reactivity, for specific size ranges, for specific coating types or structures, and in relevant suspension media), including those to characterize reactive surface area, nanometer and subnanometer surface features of ENMs, and adsorbed molecules and macromolecules on ENMs.*
- *Develop standard reactivity measures and protocols for ENMs, including a standardized approach for measuring the sensitivity of methods to important variables (for example, pH, ionic strength, organic matter, and biomacromolecules).*

Detection and Characterization of Nanomaterials in Complex Biologic and Environmental Samples

Chemical and physical information on ENMs in environmental and biologic matrices is needed. Many existing analytic techniques from material science and other disciplines are applicable to ENMs, but their use in measuring and characterizing low concentrations and heterogeneous matrices will require additional development or in some cases, development of completely new approaches. A recent review by von der Kammer et al. (2012) summarizes many of the analytic tools and research needs for detecting and characterizing ENMs in environmental and biologic matrices.

There are few analytic tools that can be used to quantify and characterize ENMs in situ (for example, in air, soil, or sediment samples), in vitro (for example, in cells or tissues), or in vivo at the low concentrations expected for most nanomaterials (in the low parts-per-billion to low parts-per-trillion range) (Hassellöv et al. 2008; Gottschalk et al. 2009; Tiede et al. 2009; von der Kammer et al. 2012). Some examples include radiolabeled materials (Hong et al. 2009; Gibson et al. 2011; Peterson et al. 2008); fluorescence (Schierz et al. 2010); mass spectrometry (MS) and single particle MS techniques (von der Kammer et al. 2012); spatially resolved X-ray analyses (von der Kammer 2012); and differential mobility analysis (Morawska et al. 2009), a well developed technique used to quantify the number and size distribution of nanoparticles in air.

Because of the lack of analytic tools, relationships between properties of materials measured ex situ (for example, nanoparticle size by TEM) and their in situ or in vivo behaviors need to be inferred, and this limits our understanding of how ENMs may be affected by such processes as in situ and in vivo transformations, biodistribution, and distribution in environmental samples.

Tools for quantifying and characterizing ENMs in the environment or in organisms typically have either a broad or a narrow spectrum. Broad-spectrum tools are applicable to a variety of sample types but require relatively high concentrations of materials (for example, non-spatially resolved synchrotron x-ray spectroscopy methods) and most often require removal from media to conditions that are not representative of in vivo or in situ environments (for example, microscopy). Narrow-spectrum tools are highly specific to a material (for example, near-infrared detection of single-walled carbon nanotubes (Leeuw et al. 2007) that can be detected at low material concentrations and potentially under in situ or in vivo conditions, but modification of the material may limit sensitivity. These narrow-spectrum tools must be developed at great expense for each type of nanomaterial. The variety of ENMs that need to be studied makes use of narrow-spectrum tools expensive and perhaps intractable. Detection in vivo or in situ can be difficult because of the low concentrations of materials released into an organism or the environment. Even if the material has not been transformed, detection is difficult; if it has been transformed, detection is even more difficult. Strategies and tools for detecting and tracking materials are needed. These strategies should include combinations of techniques to detect and characterize ENMs in complex matrices, and to differentiate between naturally occurring ENMs and naturally occurring nanomaterials (von der Kammer 2012).

Fluorescence is a common strategy that is used to localize materials, but more general techniques are needed for materials that are not fluorescent or for situations in which incorporation of a fluorescent tag interferes with the processes being investigated by modifying the material's surface properties. Another approach that will benefit nanotechnology-related EHS research is to label (for example, radiolabels) and track surface functional groups (coatings) that are being used on ENMs; however, care must be taken to ensure that the functional groups are not readily removed from the ENM by chemical or biologic reactions. Labeling approaches will need to be coupled with sensitive high-resolution methods to characterize the interactions between ENMs and the medium at the site of distribution and localization. Tracking ENMs in vivo or in situ could advance research in the field considerably, but simply tracking the presence of ENMs in these systems is not sufficient to correlate their properties with their behaviors. Methods also are needed to characterize the surface properties of ENMs in situ.

Quantifying the number and distribution of particle sizes in air samples using differential mobility analyzers (DMAs) is a well established technique (Ehara and Sakurai 2010). A DMA can quantify number concentration and size distributions, but used in isolation, it cannot determine chemical composition or surface area concentration. Further, it cannot distinguish between airborne

ENMs and naturally occurring or incidental nanomaterials. However, a DMA coupled with a single aerosol mass spectrometer can provide chemical speciation of airborne particles, and potentially can distinguish between ENMs and naturally occurring nanoparticles if the naturally occurring nanoparticles have a consistent chemical composition that is distinct from that of naturally occurring nanomaterials (Smith et al. 2010; Zhao et al. 2010). Because of the likelihood of human exposure to nanomaterials in manufacturing environments, further development of instrumentation that measures chemical composition, aggregation state, and distinguishes ENMs from naturally occurring nanomaterials in air samples is needed (for example, Zhao et al. 2010; Bzdek et al. 2011).

As discussed in Chapter 3, ENMs will be transformed in the environment (for example, by aggregation, oxidation, sulfidation, or adsorption of macromolecules). These transformations will affect distribution of ENMs in the environment or an organism. These modifications may also make their detection difficult (von der Kammer et al. 2012). Methods and tools are needed for assessing the transformation of ENMs in situ (for example, in soils, sediments, or treatment-plant effluent), in vitro (for example, in cells or tissue), and in vivo (for example, in rats).

Research Needs for Detection and Characterization of Nanomaterials in Complex Biologic and Environmental Samples

- *Develop model ENMs that can be tracked without introduction of experimental artifacts in exposure and toxicity studies.*
- *Develop analytic tools and processes that can detect ENMs at low (relevant) concentrations in situ or in vivo, followed by tools to track and characterize ENM properties (for example, reactivity, reactive surface area, nanometer and subnanometer surface features, aggregation-agglomeration, and adsorption of organic macromolecules) in situ or in vivo.*
- *Develop tools and processes to assess the rate and degree of transformation of ENMs in vivo or in situ, especially alteration of surface properties of ENMs due to adsorption of proteins and lipids (corona formation) and NOM.*

STANDARDIZED EXPERIMENTAL PROTOCOLS FOR NANOTECHNOLOGY-RELATED ENVIRONMENTAL, HEALTH, AND SAFETY RESEARCH

Development of New Protocols or Modification of Existing Protocols for Toxicity Testing and Determination of Population and Ecosystem Effects

A focused, coordinated research effort is needed to identify and validate existing or newly developed toxicity-testing protocols and best practices, such as dosimetrics (Teeguarden et al. 2007), for an agreed-on set of toxicity end points for ENMs (NRC 2007). The protocols would include rigorous physicochemical characterization of particle types, use of relevant cell types or cell systems (for example, air-liquid interface) to simulate relevant in vivo exposures, relevant dose-response protocols, relevant time-course protocols, and assessments of biomarkers, such as inflammatory end points, that have relevance to in vivo pathway models (for example, sustained inflammation). For ecologic-health research, a set of sensitive species will need to be identified in the risk-characterization phase. Development of ecologic and human-health test methods should also include coordinated interlaboratory testing validation for existing toxicity tests and end points and appropriate doses and dosing protocols in the case of newly developed tests and end points.

The appropriate end points for toxicity tests need to be determined. They should be determined from in vivo model pathways that are identified after inhalation, ingestion, or dermal exposure to ENMs. The end points for measurement after pulmonary exposure could include the following: reactive oxidant species; inflammatory biomarkers; cytotoxicity end points; cell proliferation, fibrosis, and hyperplastic responses; and histopathology, particularly for time points after exposure. For environmental health, validation of standard measures of alternative (non-acute) end points of exposure to ENMs (for example, growth, reproduction, behavior, or stress) should be determined. Additional data are needed from assay systems that have sublethal outcomes and on more types of nanomaterials to develop testing methods that are simple but have predictive value for hazard identification (Ankley et al. 2010).

To develop simple non-in vivo assays that will eventually allow high-throughput testing and provide results that predict in vivo effects (hazard identification), correlations need to be explored between the end points measured in vitro and the expected effects in vivo. That will require standardized and validated in vitro methods (for example, standardized cell types and exposure protocols) that represent specific, realistic exposures (including the materials used and the exposure routes), and doses and validation against results of in vivo studies. This is a critical step in realizing the benefits of high-throughput screening strategies proposed for ENMs.

Development of appropriate in vitro assays that can predict in vivo responses requires a detailed understanding of biodistribution of ENMs and the mechanistic pathways by which ENMs exert a toxic effect on a specific organ. Research is needed to elucidate those toxicity mechanisms for representative organisms, considering appropriate dosimetry (see above) and well-characterized ENMs, so that ENM properties can be correlated with mechanisms of injury.

Genomic tools may generate important hypotheses regarding toxicity mechanisms and may be useful for grouping nanomaterials by expected response on the basis of their properties, as has been observed in several studies with well-characterized chemicals (Bartosiewicz et al. 2001a,b; Hamadeh et al. 2002; Klaper and Thomas 2004; Dondero et al. 2011). However, in vivo data are needed to validate the genomic data with organism responses. Although genomics tools are available, research is needed to determine how much and what type of gene or protein expression changes will result in long-term effects of ENM exposures. Gene and protein expression measured in vitro with these tools must be correlated with measured effects in vivo.

The protocols for assessing ecotoxicity include those for assessing human toxicity but should also include protocols for predicting sensitive species and effects on communities and ecosystems if they are to be useful for risk assessment (Ankley et al. 2010). Those include effects on interactions among species, species community assemblages, biodiversity, and ecosystem function. There is no suite of standard tests for assessing community and ecosystem effects of chronic exposure to ENMs. That limitation is not peculiar to ENMs and presents a serious challenge to the modeling of ecologic effects.

Research Needs for Development of New Protocols or Modification of Existing Protocols for Toxicity Testing

- *Develop new standard toxicity-testing protocols or modify existing protocols for ENMs to include relevant cell types and organisms, appropriate dosimetrics, and appropriate toxicity end points (for example, chronic-toxicity end points) and validate those protocols.*
- *Identify and validate toxicity-pathway models and mechanisms to correlate in vitro end points with in vivo responses.*
- *Improve the interpretability of genomic tools by determining how gene expression and protein expression are related to ENM toxicity and mechanisms.*

Research Needs for Development of New Protocols or Modification of Existing Protocols for Determination of Population and Ecosystem Effects

- *Develop and validate a suite of standard tests that can indicate the potential for population or ecosystem effects of chronic ENM exposure on specific organisms.*
- *Develop methods for understanding ecosystem effects (that is, effects on systems of systems) that result from indirect effects of nanomaterials, such as carbon and nitrogen cycling.*

Development of New Protocols or Modification of Existing Protocols for Exposure Assessment

Exposure assessment and modeling (discussed later) will require information about sources, transport, transformations, persistence, and bioavailability of ENMs released into the environment (Johnston et al. 2010). Standard testing protocols need to be conducted to determine the properties that influence transport, transformation, persistence, and bioavailability. The protocols need to be assessed and validated with a variety of ENM types and classes and under an array of environmental conditions (for example, freshwater, seawater, terrestrial, and groundwater environments). Although it is desirable for the protocols to be applicable to a wide variety of ENMs that have differing properties and to various environmental conditions, it may not be practical, given the different environmental conditions that must be considered. Ideally, the protocols would be readily adaptable to new material properties as they are introduced.

Environmental transport will be affected by attachment of ENMs to themselves (aggregation and agglomeration) or to inorganic minerals, organic carbon, or organisms (for example, bacteria or plant roots). A variety of methods are available to measure attachment of ENMs to surfaces in the environment, including studies of column deposition and analyses using quartz crystal microbalance (Saleh et al. 2008; Petosa et al. 2010). Because of the array of variables influencing attachment of nanomaterials to environmental surfaces including physical characteristics of the nanomaterial, the properties of the environmental surface, and solution conditions, it is extremely difficult to compare attachment coefficients reported in the literature for different nanomaterials and surfaces. Standardized protocols for measuring and reporting the attachment coefficients should be developed so that the many studies of ENM attachment to environmental surfaces can be used to correlate the properties of ENMs with their propensity to attach to environmental and biologic surfaces.

ENMs released into the environment undergo a variety of transformations depending on the environmental conditions (for example, redox state, presence of NOM, or available sulfide) or biologic conditions (for example, biologic fluid and cell type) (Wiesner at al. 2006; Metz et al. 2009; Bottero and Wiesner 2010). A few types of transformations are likely to have the greatest influence on exposure potential—for example, aggregation, dissolution, adsorption of NOM or biomacromolecules, sulfidation, photodegradation, and biodegradation. Standard protocols need to be developed to "weather" or "age" ENMs in environmental and biologic media and to determine the rate and extent of transformation expected as a function of environmental and ENM properties. The aging procedures should consider expected transformations for ENMs, including dissolution, sulfidation, oxidation-reduction, adsorption of organic matter and biomacromolecules, biodegradation, and photodegradation. Relevant environmental media that should be considered include wastewater-treatment plant biosolids, terrestrial environments in which biosolids are used as fertilizer, sediment, and physiologic buffers. The most stable or metastable forms of the transformed materials and the persistence of the transformed materials should be determined. Protocols for characterizing and reporting the properties and persistence of transformed ENMs should be established to permit comparisons among studies.

Many transformations of ENMs will not produce a thermodynamic equilibrium with their surroundings, including such adsorbed macromolecules as proteins and higher-molecular-weight polymeric coatings (Casals et al. 2010), because these transformations are kinetically controlled and path-dependent. A better understanding of the kinetics of ENM transformations is needed, especially understanding of the rates of displacement of adsorbed macromolecules, for example, proteins displacing a polyethylene glycol coating on an ENM or various proteins displacing each other in plasma (Lynch et al. 2007; Walczyk et al. 2010; Monopoli et al. 2011).

ENMs may be present in environmental media at very low concentrations, so acute toxicity will not be a primary concern. However, organisms may be exposed to ENMs at low concentrations for long periods. Thus, the bioavailability of ENMs and potential long-term ecologic effects need investigation. Many factors may influence bioavailability of ENMs, including environmental conditions, physical and chemical properties of transformed ENMs, exposure routes, and timing of exposure during an organism's lifespan. Protocols are needed to assess the bioavailability of ENMs in a variety of exposure scenarios and at realistic exposure concentrations. Initially, these protocols should be developed for an agreed-on suite of indicator organisms and soil, sediment, and water-column exposures. Other organisms can be added as needed.

Standard protocols and analytic methods that measure number, surface area, and mass concentration are also needed to assess direct human exposures to ENMs via inhalation. For example, measurements from traditional mass-based exposure assessment, such as $PM_{2.5}$, are not necessarily correlated with nanoparticle number concentrations (Jeong et al. 2004). There are two distinct needs. First, personal exposure monitors are needed to collect data for occupational and epidemiology studies. Second, exposure assessment methods that are easier to operate and that can be used to support regulatory decision-making should be developed.

Research Needs for Development of New Protocols or Modification of Existing Protocols for Exposure Assessment

- *Develop and validate standard protocols for measuring and reporting attachment of ENMs to biologic and environmental surfaces.*
- *Establish protocols that can be applied to pristine ENMs to identify and classify their stability in the environment. Develop protocols to provide "weathered," transformed materials for study (for example, in transport and toxicity studies).*
- *Develop and validate protocols to assess bioavailability of ENMs to specific indicator organisms identified in a site-specific risk-assessment model.*
- *Develop standard protocols and analytic methods that measure number, surface area, and mass concentration to assess human inhalation exposures to ENMs.*

EXPOSURE MODELING

The variety of ENM types and properties necessitates the development and use of models to predict exposure to ENMs and the effects of exposure. This will include models of exposure assessment, bioavailability, mechanistic toxicity, biodistribution, and dosimetry. Proper problem formulation (see Chapter 2) is essential for the successful design and application of the models.

Risks associated with ENMs will depend on the level and time course of exposure and on the properties of ENMs to which an organism is exposed. Exposure models, whether screening-level models, such as the Environmental Protection Agency Exposure Fast Assessment Screening Tool (EPA 2010a,b), or more detailed models, such as Total Risk Integrated Methodology (TRIM FaTE, EPA 2010c), require information regarding ENM sources, transport, transformations, fate, and bioavailability (Johnston et al. 2010). Each of those has considerable uncertainty, and in many cases the tools needed to characterize them for measuring or monitoring exposure potential are lacking. Specific research needs to decrease uncertainty in exposure models are described below.

Sources of Engineered Nanomaterials in the Environment

The expected concentration of ENMs in environmental compartments cannot be predicted without reasonable estimates of the quantity of ENMs released and an understanding of the medium into which they are released (for example, air or water), their physical and chemical properties, and the spatial and temporal distribution of the releases.

The committee's conceptual framework for understanding EHS aspects of ENMs (Figure 2-1) uses an integrated value-chain and life-cycle construct as the basis of understanding of the potential for and nature of releases of and exposures to nanomaterials. That framework entails the conduct of research to identify and elucidate in considerable detail a full or at least representative array of nanomaterial value chains and life cycles. These steps also are critical to evaluating the plausibility and potential severity of risks associated with ENMs released to the environment (Chapter 2).

Many of the tools needed to predict exposure to ENMs will take time to construct. Market research is a valuable tool that can be used now to identify ENMs that are most likely to be developed in the near term. Market analysis can be used to identify key market sectors; to determine the types of nanomaterials, nanoenabled intermediates, and end products that will probably enter the market in the near term; and to estimate the volumes of materials produced, imported, and exported for specific regions. Together, those data can be used to estimate the ENMs that have the greatest potential for human and environmental exposure.

The value chain for ENMs includes many steps: synthesis, packaging, handling, shipping, and finally incorporation into intermediate and final onsumer and industrial products. Predicting human exposure to ENMs and releases into the environment requires a better understanding of those processes, including

- Manufacturing processes and handling practices.
- Postmanufacture processing.
- Storage, distribution, and transport.
- Numbers of workers involved and the nature of their activities.

- Wastes and byproducts generated and how they are managed.
- Miscellaneous routine and nonroutine activities, such as maintenance and cleaning of production equipment, upsets, and disruptions or accidents.

The potential for exposure to ENMs and their release into the environment will be influenced by the type of material; the type of application (industrial, commercial, or consumer); intended uses and reasonably expected unintended uses, including the potential to become airborne; and the potential for accidental releases over a product's life cycle. Market research can assess the potential use scenarios for specific ENMs to determine whether they are likely to be released into the environment or result in unintended human exposure. Other considerations for release potential include

- User habits and practices (for example, frequency and duration of use or misuse).
- Ancillary activities (for example, maintenance and repair).
- Wear, deterioration, and aging (for example, breakdown of exposed coatings).

End-of-use considerations include disposal, recycling of ENMs within new materials, and reuse of materials. All those present different potentials for exposure and release into the environment, and each should be determined for the types of ENM-enabled products expected on the market and their potential for release.

Releases should be characterized along the value chain of the materials to determine the most likely environmental compartments and locations to be affected and to estimate the expected concentration of ENMs in those media. A framework for determining release inventories from identified primary sources of ENMs (for example, wastewater-treatment plant effluent, biosolids, and stack emissions) should be developed. Methods for quantifying and characterizing ENMs in each source stream should be developed to validate models that have been constructed to predict ENM environmental releases. The models should be informed by large-scale efforts, such as the Nanomaterial Registry that is being supported by the National Institute of Biomedical Imaging and Bioengineering, the National Institute of Environmental Health Sciences, and the National Cancer Institute (Ostraat 2011) to collect information on nanomaterials. Additional models will need to be developed for specific release scenarios.

Research Needs for Assessing Sources of Engineered Nanomaterials in the Environment

- *Assess current and future markets for nanotechnology, identifying nanomaterials, intermediate materials, and products made with nanomaterials and nanoenabled end products that are on the market, are under development, or can be expected to emerge over the next 5-10 years.*

- *Identify the processes used to manufacture and distribute nanomaterials and nanoenabled intermediate and end products.*
- *Determine and categorize nanomaterial applications, product uses, and end-of-life scenarios for ENMs.*
- *Determine and categorize potential for releases of and exposures to ENMs.*
- *Develop models for predicting releases of ENMs into the environment along the material's life cycle and value chain.*

Transport, Transformation, and Persistence of Engineered Nanomaterials

Once released into the environment, ENMs will be transported away from the source and distribute among the environmental compartments (for example, air, water, soil, sediment, and biota). Several factors probably most influence ENM transport and fate, including the location of the release, aggregation or disaggregation, and attachment to environmental surfaces. To aid in near-term risk management of ENMs in the environment, the transport components of exposure models must be modified to include processes relevant to ENMs that are expected to behave as particles rather than as molecules. The effects of attached macromolecules or ENM matrix components, and especially NOM, on ENM aggregation and deposition need to be considered. The interrelationship between ENM properties and the media in which ENMs are dispersed (for example, freshwater, seawater, and wastewater) need to be included. Exposure models need to be updated to improve estimation of the dilution behavior of ENMs released into the environment, because the choice of assumptions about dilution will influence the prediction of ENM concentrations by the models. The models need to include aggregation, deposition, and sedimentation. A set of standardized test methods (described previously) is needed to identify the ENM properties that best predict their transport in the environment.

Ultimately, comprehensive transport models specific to ENMs must be developed and validated to determine transport away from sources and dilution in the environment, for example, modules specific to runoff from agriculture due to ENMs in biosolids applied to the fields; ENM dry and wet deposition from air, sedimentation, and sediment transport; and groundwater infiltration from agricultural activities.

As discussed in Chapter 3, ENM transformations in the environment may lead to products that have transport and toxicity characteristics different from those of their parent materials (Cheng et al. 2011), and that distribute differently in the environment. To incorporate transformations into exposure models, the likely transformations need to be determined for the different classes of materials, and the appropriate parameters for incorporating the transformations and persistence into exposure models also need to be determined. Metrics are needed for describing the extent of ENM transformation and the persistence of ENMs. For example, a simple half-life time for loss of ENM mass may be appropriate for ENMs that dissolve, but more complex metrics such as size, change in number concentration, or change in reactivity with time may be more appropriate for ENMs that only partially transform.

Research Needs for Assessing Transport and Fate of Engineered Nanomaterials in the Environment

- *Modify exposure models to include processes most relevant to assessing ENM distribution in the environment and human exposure (for example, aggregation, degradation rate, and dilution).*
- *Incorporate into exposure models metrics for transformation and persistence (for example, half-life time, size, change in number concentration or surface area, or change in reactivity with time).*

MODELS FOR PREDICTING HUMAN HEALTH, ORGANISMAL, AND ECOLOGIC EFFECTS

As discussed in Chapter 3, a key requirement for use of any in vitro assay as a predictive tool is that the underlying mechanisms also operate in vivo. For human health and ecosystem considerations, there are at least four generally recognized toxicity mechanisms: inflammation and oxidative stress, immunologic mechanisms, protein aggregation and misfolding, and DNA-damage mechanisms. Effects models should consider each of those rather than relying solely on oxidative stress end points. In addition, there may be other mechanisms that have yet to be identified. The sections below discuss modeling needs for assessing ecosystem effects.

A major ecotoxicology modeling goal has been the development of reliable, predictive models whereby the chemical structure of a compound can be used to predict the harm that it will cause. That is particularly challenging for ENMs given the breadth of ENM types and the lack of understanding of ENM transformations in the environment and in organisms. Toxicity mechanisms of ENMs should be a major research focus to augment the models. Predictive models are most accurate when a specific mechanism is being probed, whereas a whole-organism response in the absence of knowledge of the mechanism provides poor modeling and predictions (Schmieder et al. 2003; Ankley et al. 2010).

Developing assays that provide unbiased parameters and that elucidate different mechanisms of action will support better models. Data should include multiple end points for a chronic life-cycle assessment. Low-dose effects need to be assessed. Multiple organisms need to be considered, as do multiple pathways. To that end, more information is needed on the pathways that are disrupted in whole-organism assays. Duration of exposure needs to be considered as part of the effects model because these effects may change over time as a result of accumulation or formation of byproducts in the organism or the recovery pathways. Chronic and sublethal endpoints such as effects on interactions among species, species-community assemblages, biodiversity, and ecosystem function should be considered (Bernhardt et al. 2010).

Research Needs for Predicting Organismal, Population, and Ecosystem Effects

- *Improve development of toxicity models by measuring and reporting on more sublethal toxicity end points—including, growth, behavior, reproduction, development, and metabolism—in more ENMs that have specific core and surface properties.*
- *Develop models to link the biochemical pathway responses to ENMs and ENM properties (beyond oxidative stress) to adverse outcomes.*
- *Develop models that can predict organismal, population, and ecosystem effects of exposure to ENMs through the collection of more data on community and ecosystem effects and the determination of pathways for adverse outcomes (for example, effects on survival and reproduction).*

EXPOSURE TO DOSE MODELS

Models can be developed to predict the concentrations of ENMs in various environmental media (for example, air, water, soil, and sediment) and thus to predict exposure. Dosimetry is needed to correlate measurable exposure concentrations with adverse outcomes observed in exposed organisms. Dosimetry is organism-dependent and therefore difficult to predict without organism-specific models; however, decisions about acceptable concentrations of ENMs in environmental media or acceptable worker exposure will require understanding of relationships between ENM properties, environmental conditions, routes of exposure, and doses to an organism. These considerations extend to models of both human and environmental exposures. Ultimately, dosimetry models must be developed to predict dose from exposure concentrations.

Existing exposure models need to be updated to be applied to ENMs. As discussed in Chapter 3, human inhalation exposure to ENMs is an important route of exposure to humans. Following uptake of ENMs, they may be distributed throughout an organism and reside in locations distal to the initial exposure site. Exposure models, such as the Multiple Pass Particle Dosimetry model need to be updated to include the effects of different particle shapes and translocation from the deposition site in the respiratory tract to other organs. Biodistribution models are essential for identifying specific organs that may be targeted, injury mechanisms, and the toxicity assays that best represent the exposures and mechanisms of toxicity. A repository of validated reference or benchmark nanomaterials, whose surfaces can be modified, should be established to simulate the effects of diverse surface properties on their disposition; that will help to assess biodistribution in animal models after exposure by different routes. Specifically, the influence of exposure concentration on biodistribution should be investigated. Establishing a comprehensive biodistribution model—including uptake, translocation, and elimination pathways and mechanisms—will be an important input for bioinformatics.

Because ENMs will probably be present in environmental media at very low concentrations, acute toxicity may not be a primary concern. However, organisms may be exposed to ENMs at low concentrations for long periods. A long-term goal is to predict ENM bioavailability to and bioconcentration in relevant organisms on the basis of properties of the ENMs and their transformation products. That is a formidable task because bioavailability and bioconcentration will probably depend in part on ENM properties, route of exposure, and dispersion media. Thus, the ENM properties and environmental conditions affecting bioavailability and bioconcentration of ENMs in long-term exposure need to be determined and incorporated into exposure models.

Research Needs for Developing Exposure-to-Dose Models

- *Update dosimetry models by determining deposition efficiency and underlying mechanisms of inhaled ENMs throughout the respiratory tract depending on shape, surface properties, and agglomeration of the ENMs.*
- *Support development of biodistribution models by identifying reference or benchmark ENMs with appropriate labels to serve as models for assessing disposition after exposure by different routes (in air, water, and food).*
- *Develop models to predict bioavailability of ENMs on the basis of their properties, routes of exposure, and exposure media.*

Model Uncertainty

The uncertainty surrounding environmental exposure to and effects of ENMs is substantial given the paucity of data on the behavior of ENMs in the environment and the measured quantities of ENMs in environmental media. Research efforts may never decrease inherent uncertainties to the point where deterministic models are possible. Because of the high degree of uncertainty regarding the ENM properties that influence fate and exposure potential, the use of probabilistic exposure-assessment models is needed. That is not unlike the risk-informed decision-making (RIDM) approach used by the Nuclear Regulatory Commission (Vietti-Cook 1999). RIDM, as defined by the commission, is an approach to regulatory decision-making in which insights from probabilistic risk assessment are considered with other engineering insights. The approach asks simply, What can go wrong? How likely is it? What are the consequences? (Vietti-Cook 1999).

A long-term goal is to decrease uncertainty in results of exposure and effects models to better support decision-making with regard to nanotechnology-related EHS risks. Approaches to identifying research that will provide the greatest reduction in uncertainty in exposure modeling may benefit from a value-of-information analysis as described in NRC (2009).

Models used to predict exposure to and effects of ENMs must be capable of incorporating the uncertainties of the input parameter values to ensure that model results are expressed in the context of the uncertainties. Stakeholders and regulators who make risk-management decisions regarding ENMs will need to become familiar with probabilistic models and interpretation of their results.

Research Needs for Incorporating Uncertainty into Models and Reporting Uncertainty in Exposure Modeling

- *Identify key uncertainties surrounding exposure-assessment models and estimate the ranges of the uncertainties.*
- *Incorporate uncertainty into models that are being developed to estimate sources, model transport and fate, bioavailability, and effects.*

INFORMATICS

Predicting the potential effects of ENMs on the basis of their chemical properties will require a long-term funding commitment to nanotechnology-related EHS research and a highly coordinated research effort with standardized data collection and warehousing that allow mining of a highly diverse set of data (and metadata) types and formats. It will require an appropriate informatics infrastructure that addresses the primary research needs discussed above and that supports more efficient approaches to increasing collaboration and data-sharing among the disciplines involved in nanotechnology research, development, translation, and regulation.

The previous sections of this chapter addressed major research needs resulting from a lack of information on

- The availability of reference nanomaterials.
- The characterization of nanomaterials as reported in the literature and in publicly available databases.
- Errors and uncertainty in the methods and protocols used to produce the data.
- The sensitivity of the methods to variations in experimental apparatus, materials, and procedures.
- Errors and uncertainty in and the sensitivity of the models used to evaluate ENM structure-property relationships, to predict their behavior and effects, and to evaluate and manage risks.

Although the nanotechnology community is distinguished by the wide spectrum of disciplines involved, the lists of research needs presented here could be shared to some degree with any emerging technology that, by definition, is beginning to produce new and as yet insufficiently tested products. In particular, errors and uncertainty in and the sensitivity of the methods and models used should be established because they are needed for improving product and process design and for predicting and managing risk. The current need to manage life-cycle risk posed by nanotechnology products presents an opportunity to establish an infrastructure that can serve as an exemplar of emerging technologies.

Nanotechnology and nanoscience span basic and applied research and translational and regulatory domains, including manufacturing, process control, and human and environmental health. The timescales for developing new standard methods from available research protocols is normally decades. For nanotechnology, the timescales must be compressed dramatically because of the introduction of so many new products and components that contain nanomaterials. Sharing information among these broad and diverse domains requires attention to ensure that the information communicated is interpreted properly and unambiguously and that the transmission is accomplished within the appropriate timescales and resources. Although modern information technology offers unprecedented tools and applications for rapid communication, data storage, transmission of semantic content, and support for collaborative enterprises, technology alone cannot provide the needed solutions. New legal, social, and cultural approaches and mechanisms will be required to permit more comprehensive and time-appropriate information-sharing as nanotechnology products continue to proliferate.

In 2004, international efforts to standardize nanotechnology were initiated and emphasized the need for standard analytic methods and protocols for characterizing the physicochemical properties of nanomaterials and their activity in in vitro and in vivo studies and the need for common terminologies to harmonize communication among different disciplines and stakeholders (OECD 2010; IANH 2011). Additional efforts followed quickly, including development of standards for the minimum information required for characterization of nanomaterials (Aitken et al. 2009; Card and Magnuson 2009; MINChar 2009; Boverhoff and David 2010), for taxonomies and ontologies to augment metadata with semantic content about nanomaterial properties (Gordon and Sagman 2003; Kozaki et al. 2011; Thomas et al. 2011), and for harmonization of formats for data-sharing (ASTM 2010; Klemm et al. 2010).

The need for new approaches to provide the requisite informatics infrastructure are described here with the need for information-sharing to provide an understanding of the collaboration timescales, information technologies, and resources required. Approaches and informatics requirements are described as they are related to method development and validation, model development and validation, and data management and data-sharing. Research priorities to develop the knowledge base and data-sharing capabilities are presented in Chapter 5. Means of enhancing collaboration necessary for implementation of an informatics infrastructure are discussed in Chapter 6.

Method Development and Validation

Standard methods and instrumentation developed for nanomaterials research must be adapted to process control for manufacturing, recycling, waste processing, regulation, and remediation. In addition, documentation on standard guidelines and practices is needed. Standard-method (protocol) development is difficult. The development and validation of standard methods is a long process that requires exhaustive testing of the precision, accuracy, reliability, and reproducibility of the methods and of the sensitivity of results to changes in protocol parameters, instrumentation, environment, media, and models (that is, the ruggedness and robustness of the method). Standard-development organizations (SDOs) require years to reach consensus on new standards, and testing protocols to obtain measures of their error and uncertainty through interlaboratory studies (ILSs) requires additional time and resources. Reference nanomaterials for ILSs, instrument calibration, and validation are generally not available, particularly in large, well-characterized batches with sufficient stability. In addition, shipping the reference and sample nanomaterials, cell lines, media, and materials required by a protocol and materials needed for sample preparation may require environmentally controlled shipping containers and data-logging devices to ensure that the materials are not exposed to extreme conditions in transit. Gauging adherence to standard methods is also a problem because of the difficulty of providing a documented standard with sufficient detail regarding necessary additional positive or negative controls and sample preparation, and because of the lack of any data-reporting standard regarding deviations from the method. Involvement in SDOs is a voluntary activity: industry has incentives to participate in standards development, but academic participation, at least in the West, is hindered by the lack of funding and academic credit for participation in standard development (CTAC 2007; OSI 2007). Collaboration among SDOs is also lacking and is usually confined to postdevelopment harmonization of standards.

Because of those compounded difficulties in the development and dissemination of validated standard methods, recent ILSs of new protocols by the Asia Pacific Economic Forum, the American Society for Testing and Materials, and the International Alliance for NanoEHS Harmonization have shown that the reliability and reproducibility of published nanotechnology data are problematic (Hackley et al. 2009; Murashov and Howard 2011). In addition, the ability to qualify laboratories in the performance of new validated methods is extremely restricted. As a result, contract research organizations (CROs) are not available to perform the extensive characterizations required, and individual laboratories must perform the measurements themselves. That leads to variability in characterization results and in inefficient use of funding for nanotechnology-related EHS research. However, it should be noted that outsourcing of material characterization by CROs may not be effective in all cases; for example, characterization efforts must be closely linked to synthetic efforts because rapid feedback is needed to develop the best materials, and this is not conducive to outsourcing. Given these challenges, the establishment and broad dissemination of best practices to support local characterization and real-time feedback to nanomaterial producers would significantly enhance data quality and consistency, without invoking standards and certified instruments that are prohibitively expensive.

Needs in Method Development and Validation

- *Increase the amount, quality, and availability of nanomaterial characterization and effects data by accelerating the development of validated standard methods and interlaboratory study of the methods.*
- *Provide greater detail of the methods for sample preparation and required controls for specific nanomaterials.*
- *Remove the barriers to qualifying CROs to perform the needed extensive characterizations of nanomaterials by accelerating the development of these more detailed and validated standard methods and by ensuring environmental control of method materials and nanomaterial samples during shipping.*

Those general needs are of two types: (1) needs that can be addressed through traditional informatics methods, applications, and tools and (2) needs that require new approaches and mechanisms to ensure collaborative participation.

Informatics needs that require enhancement of traditional method development and validation procedures include the ability to collect, organize, curate, and share data on

- *The methods used, including links to standard protocols.*
- *Deviations from the methods that were implemented, including additional controls or sample-preparation techniques for specific nanomaterials.*
- *Links to error, uncertainty, and sensitivity data on the methods as determined through the method-validation process and interlaboratory studies.*
- *The level of validation and expertise for laboratories using the methods.*
- *Whether minimum characterization of nanomaterial standards were met in carrying out an investigation.*
- *Links to training materials to assist laboratories in adopting new standard methods and techniques.*

Providing a similarly concise categorization of new approaches and mechanisms to accelerate development and validation of new methods is difficult primarily because current practices are deeply ingrained in different disciplines and communities and shortening the timeframes for transitioning from pure hypothesis-driven research methods to a translational and regulatory method-validation framework will necessarily disrupt current practices. In short, cultural barriers need to be overcome, and the more tractable technical barriers to implementing a new informatics infrastructure need to be addressed. However, documented examples of how to overcome these barriers through digital communication are available (Goble et al. 2010; Tan et al. 2010).

Approaches needed for new collaborative mechanisms for method development and validation include the ability to provide a framework and additional incentives for

- *Broader participation in method development and validation, including interlaboratory studies, particularly among academics but also interinstitution, interagency, and international collaboration, including that of regulatory bodies, metrology institutes, and national laboratories.*
- *Participation in responding to the specific informatics needs delineated previously while ensuring that data curation is performed by those with the most expertise in evaluating the quality of each particular dataset, that data rights are determined by the owners of the data, and that user requirements are provided by all user communities, including nongovernment organizations and the public.*
- *Increased collaboration and harmonization among standard-development organizations, contract research organizations, nanomaterial providers, and organizations that conduct interlaboratory studies to develop methods to make the most efficient use of the available pool of experts and the best available materials and to minimize the need for later efforts in harmonization among standards produced by different organizations.*

An implementation scenario for development of methods and protocols is described in Appendix B.

Model Development and Validation

The previous sections described the need for improving model development and validation efforts with regard to ENMs—models for predicting environmental release, transport and fate, exposures and their relation to dose, and human health, organismal, and ecologic effects—and for risk assessment and determination of quantitative structure-activity relationships. Each of the models incorporates submodels of different types—models of in situ environments; cellular, tissue, organ, system, organism, and ecosystem models; and models of ENM structure that use basic descriptors, detailed molecular structures, or both and that use different numerical and statistical applications and tools. Although it may appear that the requirements for model development and validation are similar to those for method development and validation, there is a striking difference: the models are in a computer-usable form. That difference opens the possibility of more rapid, collaborative development and validation of models than can be achieved for methods. It is an encouraging distinction inasmuch as the greater use of predictive models in conjunction with experimental results can help in discovering underlying mechanisms.

The Worldwide Protein Data Bank (wwPDB) is a relevant example. The wwPDB was originally designed to serve both as a repository for protein molecular structures and as a collaborative mechanism to curate and validate structural models and to improve predictive models of protein structure, conformation, binding, and activity in different environments (Berman et al. 2007). Although today the wwPDB is primarily a repository, the use of the structural models continues to be an important tool for improving and validating predictive models. A similar worldwide repository for molecular structures of ENMs that could validate predictive models is needed.

Extending the PDB of nanomaterials concept to include a database of predictive models would offer some desirable alternatives to current procedures. (A discussion concerning the challenges of developing such a database is described in Appendix B, under the heading Development and Validation Scenario.) Scientific collaboration in model development through publications imposes multiyear delays: a model is first developed or improved; new results are obtained and verified with the model; a publication is written, published, and read, and its findings are corroborated by other researchers; and a new improvement is made in the model. In an alternative scenario, in light of the PDB of nanomaterials concept, the initial researcher could make the predictive model available on the Web with all the files, run-time parameters, and test scripts necessary to duplicate the result—and save years. The technology to support such accelerated collaboration is widely available, and credit can be given to both the model developer and the person who improves the model. The advantages of faster, open access to scientific results have been well documented in physics (for example, Gentil-Beccot et al. 2009), and open-access collaborative networks have proven to be very effective in advancing biomedical research and translating the results to the clinic (Derry et al. 2011). In fact, collaborative code development is recognized in industry as the preferred means of producing reliable computer applications and is used in existing collaborations. Adapting the technique to model and submodel development and validation on an international scale would dramatically shorten the timescale for model improvement. Such model-development practices would also enhance scientist-to-scientist collaboration on problems of mutual interest with few operational resources and provide more facile options for leveraging and sharing intellectual property through knowledge-sharing networks and sites such as Creative Commons and IC Tomorrow. The magnitude of the changes occurring in managing and using intellectual property are reflected in the Creative Industries Knowledge Transfer Network (2011) report, p. 30:

The business model evidence…"indicates that the creative industries are being pushed increasingly towards realizing value from IP [intellectual property] sharing. The precedent of open source suggests that the highest value will be placed on IP that delivers scale through widespread adoption and use".

Informatics Needs for Model Development and Validation

- *An informatics portal similar to the wwPDB is needed to archive, organize, curate, validate, and share structural models of nanomaterials and their surface coatings, both pristine and transformed, for collaborative use internationally.*
- *An extension of the database is needed to archive, organize, curate, validate, and share the predictive and probabilistic models and submodels to accelerate their development and use and to augment and complement experimental techniques.*
- *New mechanisms are needed to aid in implementing the required collaborative databases for structural models of ENMs and for development of predictive and probabilistic models.*

An implementation scenario for development of predictive and risk models is discussed in Appendix B.

Data-Sharing

The preceding discussions highlighted the need for sharing data from specific protocols; for example, for characterization of pristine and transformed ENMs in complex samples, for toxicity tests and ecosystem and population effects, and for methods for exposure assessment and for characterizing transformed and weathered ENMs. The variety of protocols reinforces the need for data-sharing among diverse disciplines that use different techniques and practices. It is important to provide data-sharing techniques that provide the scientific data requested and that describe both the data and the ENMs with sufficient detail to track which manufactured lot of a material the ENM samples were taken from to account for lot-to-lot variability. The informatics system should use specific nomenclatures and terms that have agreed-on definitions. The same requirements are appropriate for supporting model development and validation.

Experience with today's search engines, however, illustrates the lack of specificity that is achieved when only search terms are used—for example, the millions of "hits" that might have to be sifted through to extract the desired information. Adding semantic content about the meanings and relationships of the search terms is the aim of the Semantic Web, sponsored by the World Wide Web Consortium of informatics systems providers and users (W3C Semantic Web 2011). Providing machine-usable logical relationships about and among search terms allows a search engine to reduce drastically the number of false "hits." Ontologies improve the use of search terms by supplying definitions for each term and explicit logical relationships among the terms specified so that they can be interpreted and used by a computer. The effort to generate and maintain an ontology is usually supported by a community that agrees on the concept definitions and logical relationships. Ontologies can be modified or extended to accommodate different communities, and methods exist to map relationships between terms in different ontologies, allowing any (mapped) ontology to be used to mine data in a dataset organized by terms of a different ontology. Currently, organizations involved in nanotechnology, both nationally and internationally, support Web sites and portals to provide information and analyzed data. Very few portals offer raw data that may be stored locally in the laboratory generating the data. Most of the portals use their own systems and offer their information through current search engines. Attempting to harmonize the search terms, data formats, curation levels, and security for these portals would entail a large enterprise that would be prone to failure because the databases may be linked to back-office applications that would be difficult to modify. Moving all the data to a central site has been tried and is usually very difficult because of issues involving data rights and security and the need to agree on common formats, procedures, and rules for governance.

Informatics Needs for Data-Sharing

- *Use existing pilots to demonstrate the capability to federate the different sites through the use of semantic web technologies, including ontology development to enable data curation by experts in the data, and access control by the owners of the data.*

Using that method would allow international entities to come together on an equal footing to craft a short-term solution for collaborative protocol and model development. A modest effort would be needed to demonstrate current capability as a pilot project for use in an interoperable system to establish user requirements relevant to the entire community, and initial efforts are being undertaken within the nanotechnology informatics community (InterNano 2011).

- *Use and modify existing ontologies and semantic web applications, perhaps in collaboration with search-engine providers, to develop an automatic ontology "crawler" to update mappings among the ontologies used or adopted by the collaborating partners.*

Barriers to Informatics

Successful implementation of the informatics strategy described in this chapter—including developing ontologies, data-sharing, and community model development—requires appropriate datasets as inputs. The emerging field of nanoinformatics, in contrast with the more fully developed field of bioinformatics, faces some specific challenges. Biopolymers are often discrete structures or sequences, whereas nanomaterials typically exhibit a dispersion of sizes, compositions, and surface coatings. Such dispersions are difficult to define and reduce to the precise code needed for informatics. Second, given the wide array of nanomaterial types, structural information for different classes of materials will be based on different sets of analytic measurements that make direct comparisons difficult. Because no one measurement can describe a nanomaterial completely, an informatics approach will need to synthesize the information from multiple techniques to describe the material. Given the number of gaps in the data on the nanomaterials described in the literature, most materials are now incompletely described and will probably remain so unless incentives are developed to characterize them. Finally, whereas biopolymers can be readily described by reference to their primary sequence and a series of letter codes or by a defined three-dimensional structure determined with x-ray crystallography, the different types of measurements (for example, images, histograms, optical spectra, and elemental composition) that are used to define nanomaterials are difficult to reduce to code.

Those complexities will result in barriers to the development of nanoinformatics unless they are addressed through close interaction with the scientists who are producing and characterizing the new nanomaterials. One barrier is the relatively onerous process of data entry for nanomaterials. If the materials cannot be described as single structures or sequences, as is possible for biopolymers, describing their dispersity makes the process more time-consuming. In addition, uploading raw data that are in a wide array of nonstandard formats presents a barrier to those who might contribute to the database of materials. But it is important to have access to the raw data because producing a numerical descriptor from them often involves considerable interpretation.

Who will generate the data for informatics, and what are the incentives for them to participate? From one perspective, the information used to populate the databases for nanoinformatics efforts will be developed by specialists using standard protocols and working with defined reference materials. That approach is relatively slow—working with one painstakingly produced and characterized material at a time. More rapid progress could be made if information on all materials produced and characterized could be captured in the databases regardless of who produces the materials. The presence of such data would encourage biologists and toxicologists to study the materials, but what is the incentive for the nanomaterials chemist to contribute this information?

Recommendations for addressing barriers to informatics for nanomaterials:

Provide incentives to nanomaterials innovators to characterize and report sufficient analytic data to define materials for comparison with other materials, including error, uncertainty and sensitivity data. For example,

- *Journals could require the data for publication.*
- *Agencies could make collecting and sharing the data conditions for funding, perhaps through National Science Foundation data-management plans (see discussion in Chapter 6) or more specifically in nanotoxicology grants.*

Recognize the challenges in comparing data. For example, size and size distributions from transmission electron microscopy are not all the same. They are not the raw data; the image is. Although it is necessary to obtain a representative sample from a set of images, new standard methods for nanomaterials (for example, NIST/NCL 2010; ASTM 2011) reference best practices to control image selection bias (for example, Allen 1996; Jillavenkatesa et al. 2001), and evaluations of bias in instrumentation software for defining particle boundaries are being considered. It is difficult to develop structure-activity relationships when the structures are not concretely defined.

Reduce barriers to nanomaterial innovators contributing to databases by engaging with them, understanding the complexities, and finding solutions that reduce the barriers.

Provide incentives for companies to provide information on nanomaterials that they have pioneered. This will require finding creative ways to protect intellectual property.

Work toward a model, such as the PDB of nanomaterials concept, but engage nanomaterial-synthesis experts at the beginning to identify and find solutions to obstacles.

TABLE 4-1 Summary of Research Needs Identified in Chapter As Mapped to the Tools

MATERIALS
Well-characterized materials are needed, including: reference materials of varied size, shape, aspect ratio, surface charge, and surface functionality for testing; "real-world" materials for testing; "weathered" nanomaterials that are representative of those expected in vivo or in situ; materials that can be tracked (for example, for biodistribution or environmental partitioning studies) without introducing experimental artifacts in exposure and toxicity studies; and standard reference materials to use in calibrating assays and measurement tools.
METHODS
Develop and validate new or modify existing standard toxicity-testing protocols for ENMs, including relevant cell types and organisms, appropriate dosimetry and toxicity end points (for example, chronic effects), and gene and protein expression to identify and validate toxicity mechanisms, such as biodistribution of ENMs and toxicity-pathway models.
Develop methods to extrapolate and predict long-term low-dose effects from short-term high-dose effects, and validate their accuracy through blinded test methods.
Develop screening methods that can indicate the potential for bioavailability and potential for effects due to chronic ENM exposure or for indirect effects, that is, not direct toxicity from ENM exposures (for example, the effects of ENMs on carbon and nitrogen cycling).
Develop and validate standard methods for measuring and reporting attachment affinities of ENMs to biologic and environmental surfaces to facilitate assigning values to parameters in exposure models.
Develop methods to determine the reactivity and stability of ENMs in biologic and environmental samples, including standard measures for assessing and reporting reactivity (for example, generation of reactive oxygen species).

(Continued)

TABLE 4-1 Continued

Develop a standardized approach for measuring a method's sensitivity to changes in important variables (for example, pH, ionic strength, organic matter, and biomacromolecules) and standard ways to report sensitivity.
INSTRUMENTATION
Develop new instrumentation and methods for existing instrumentation to isolate subpopulations of ENMs from polydisperse samples.
Develop tools that can detect ENMs, especially at low (relevant) concentrations in situ or in vivo, followed by methods to track and characterize ENM properties (for example, reactivity, reactive surface area, nanometer and subnanometer surface features, aggregation, and adsorption of organic macromolecules). In the future, develop methods that operate unattended and monitor ENMs in the environment in different media, especially air and water.
Develop tools to assess the rate and degree of transformation of ENMs in vivo or in situ, especially specific alteration of surface properties of ENMs due to adsorption of proteins and lipids (corona formation) and NOM.
MODELS
Develop models to estimate sources of ENMs released into the environment along a material's life cycle and value chain.
Modify traditional exposure models to include processes that affect ENM distribution in the environment and influence human exposure (for example, attachment to environmental and biologic surfaces, degradation rate, and dilution) and determine how to assign values to parameters in those models.
Determine toxicity pathways for outcomes (for example, effects on survival and reproduction) that predict population effects of ENM exposure and formulate ecotoxicity models, using data on sublethal toxicity end points (including effects on growth, behavior, reproduction, development, and metabolism).
Update inhalation models to include dependence on ENM shape, surface properties, and agglomeration on deposition efficiency, and the underlying mechanisms of deposition of inhaled ENMs in the respiratory tract.
Identify pathways of elimination of ENMs after their biodistribution and accumulation in primary and secondary organs. Determine principle mechanisms of elimination as inputs into predictive bioinformatics modeling.
Identify key uncertainties and sensitivities surrounding exposure assessment and effects models, estimate the ranges of the uncertainties and sensitivities, and incorporate the uncertainties into the models.
INFORMATICS
Identify minimum characterization principles to develop standardized descriptors (that is, metadata) for ENMs that are related to their key physical material characteristics for reporting and cross-referencing data on ENM properties and effects.
Establish uniform metadata to describe ENM manufacturing and distribution processes and to correlate lot-to-lot variability of ENM properties with changes in synthesis and handling.
Develop ontologies and data formats to allow relevant data on gene and protein expression to be correlated with ENM-toxicity mechanisms.
Develop strategies for federating nanotechnology databases administered by different agencies, business entities, universities, and nongovernment organizations to allow seamless data exposure and data-sharing while protecting intellectual-property rights.
Develop new mechanisms for digital archiving and annotating and updating of methods, data, tools, and models to spur rapid and efficient formation of new targeted national and international scientific collaborations.
Develop and augment ontologies to support nanotechnology and nanoscience and in particular to develop an ontology "crawler" to aid in mapping relationships among ontologies.

REFERENCES

Aitken, R.J., P. Borm, K. Donaldson, G. Ichihara, S. Loft, F. Marano, A.D. Maynard, G. Oberdörster, H. Stamm, V. Stone, L. Tran, and H. Wallin. 2009. Nanoparticles - one word: A multiplicity of different hazards. Nanotoxicology 3(4):263-264.

Allen, T. 1996. Particle Size Measurement, Vol. 1. Powder Sampling and Particle Size Measurement, 5th Ed. London: Chapman & Hall.

Alvarez, P.J., V. Colvin, J. Lead, and V. Stone. 2009. Research priorities to advance eco-responsible nanotechnology. ACS Nano. 3(7):1616-1619.

Ankley, G.T., R.S. Bennett, R.J. Erickson, D.J. Hoff, M.W. Hornung, R.D. Johnson, D.R. Mount, J.W. Nichols, C.L. Russom, P.K. Schmieder, J.A. Serrano, J.E. Tietge, and D.L. Villeneuve. 2010. Adverse outcome pathways: A conceptual framework to support ecotoxicology research and risk assessment. Environ. Toxicol. Chem. 29(3):730-741.

ASTM. 2010. ASTM WK28974 - New Specification for a Standard File Format for the Submission and Exchange of Data on Nanomaterials and Characterizations. ASTM International, West Conshohocken, PA [online]. Available: http://www.astm.org/DATABASE.CART/WORKITEMS/WK28974.htm [accessed June 8, 2011].

ASTM. 2011. ASTM WK29480 - New Guide for Size Measurement of Nanoparticles Using Atomic Force Microscopy (AFM). ASTM International, West Conshohocken, PA [online]. Available: http://www.astm.org/DATABASE.CART/WORKITEMS/WK29480.htm [accessed Nov. 23, 2011].

Bartosiewicz, M.J., D. Jenkins, S. Penn, J. Emery, and A. Buckpitt. 2001a. Unique gene expression patterns in liver and kidney associated with exposure to chemical toxicants. J. Pharmacol. Exp. Ther. 297(3):895-905.

Bartosiewicz, M., S. Penn, and A. Buckpitt. 2001b. Applications of gene arrays in environmental toxicology: Fingerprints of gene regulation associated with cadmium chloride, benzo(a)pyrene, and trichloroethylene. Environ. Health Perspect. 109(1):71-74.

Berman, H., K. Henrick, H. Nakamura, and J.L. Markley. 2007. The worldwide Protein Data Bank (wwPDB): Ensuring a single, uniform archive of PDB data. Nucleic Acids Res. 35 (suppl. 1):D301-D303.

Bernhardt, E.S., B.P. Colman, M.F. Hochella, Jr., B.J. Cardinale, R.M. Nisbet, C.J. Richardson, and L. Yin. 2010. An ecological perspective on nanomaterial impacts in the environment. J. Environ. Qual. 39(6):1954-1965.

Bottero, J.Y., and M.R. Wiesner. 2010. Considerations in evaluating the physicochemical properties and transformations of inorganic nanoparticles in water. Nanomedicine 5(6):1009-1014.

Bouwmeester, H., I. Lynch, H.J. Marvin, K.A. Dawson, M. Berges, D. Braguer, H.J. Byrne, A. Casey, G. Chambers, M.J. Clift, G. Elia, T.F. Fernandes, L.B. Fjellsbø, P. Hatto, L. Juillerat, C. Klein, W.G. Kreyling, C. Nickel, M. Riediker, and V. Stone. 2011. Minimal analytical characterization of engineered nanomaterials needed for hazard assessment in biological matrices. Nanotoxicology 5(1):1-11.

Boverhof, D.R., and R.M. David. 2010. Nanomaterial characterization: Considerations and needs for hazard assessment and safety evaluation. Anal. Bioanal. Chem. 396(3):953-961.

Bzdek, B.R., C.A. Zordan, G.W. Luther, and M.V. Johnston. 2011. Nanoparticle chemical composition during new particle formation. Aerosol Sci. Tecnol. 45(8):1041-1048.

Card, J.W., and B.A. Magnuson. 2009. Proposed minimum characterization parameters for studies on food and food-related nanomaterials. J. Food Sci. 74(8):vi-vii.

Casals, E., T. Pfaller, A. Duschl, G.J. Oostingh, and V. Puntes. 2010. Time evolution of the nanoparticle protein corona. ACS Nano. 4(7):3623-3632.

Cheng, Y.W., L.Y. Yin, S. Lin, M. Wiesner, E. Bernhardt, and J. Liu. 2011. Toxicity reduction of polymer-stabilized silver nanoparticles by sunlight. J. Phys. Chem. C 115(11):4425-4432.

Creative Industries KTN (Knowledge Transfer Network). 2011. Beacon 10 IP & Open Source, Final report. Creative Industries Knowledge Transfer Network, University of Arts, London [online]. Available: https://connect.innovateuk.org/c/document_library/get_file?p_l_id=1342553&folderId=1812583&name=DLFE-33289.pdf [accessed Nov. 10, 2011].

CTAC (Clinical Trials Advisory Committee). 2007. First Clinical Trials Advisory Committee Meeting, January 10, 2007, Bethesda, MD. National Institutes of Health, National Cancer Institute [online]. Available: http://deainfo.nci.nih.gov/advisory/ctac/0107/10jan07mins.pdf [accessed May 25, 2011].

Derry, J., L.M. Mangravite, C. Suver, M. Furia, D. Henderson, X. Schildwachter, J. Izant, S.K. Sieberts, M.R. Kellen, and S.H. Friend. 2011. Developing predictive molecular maps of human disease through community-based modeling. Nat. Prec. 713 (April 4, 2011): doi:10.1038/npre.2011.5883.1.

Dondero, F., M. Banni, A. Negri, L. Boatti, A. Dagnino, and A. Viarengo. 2011. Interactions of a pesticide/heavy metal mixture in marine bivalves: A transcriptomic assessment. BMC Genomics 12(1):195.

Ehara, K., and H. Sakurai. 2010. Metrology of airborne and liquid-borne nanoparticles: Current status and future needs. Metrologia 47(2):S83-S90.

EPA (U.S. Environmental Protection Agency). 2010a. Exposure and Fate Assessment Screening Tool Version 2.0 (E-FAST V2.0) [online]. Available: http://www.epa.gov/opptintr/exposure/pubs/efast.htm [accessed May 8, 2011].

EPA (U.S. Environmental Protection Agency). 2010b. Interim Technical Guidance for Assessing Screening Level Environmental Fate and Transport of, and General Population, Consumer, and Environmental Exposure to Nanomaterials. U.S. Environmental Protection Agency. June 17, 2010 [online]. Available: http://www.epa.gov/opptintr/exposure/pubs/nanomaterial.pdf [accessed Apr. 23, 2011].

EPA (U.S. Environmental Protection Agency). 2010c. Total Risk Integrated Methodology (TRIM) – TRIM.FaTE [online]. Available: http://www.epa.gov/ttn/fera/trim_fate.html [accessed May 8, 2011].

Gao, X.H., L.L. Yang, J.A. Petros, F.F. Marshall, J.W. Simons, and S. Nie. 2005. In vivo molecular and cellular imaging with quantum dots. Curr. Opin. Biotechnol. 16(1):63-72.

Gentil-Beccot, A., S. Mele, and T.C. Brooks. 2009. Citing and reading behaviours in high-energy physics. How a community stopped worrying about journals and learned to love repositories. SLAC Scientific Documents No. 13693 [online]. Available: http://slac.stanford.edu/pubs/slacpubs/13500/slac-pub-13693.pdf [access Nov. 23, 2011].

Gibson, N., U. Holzwarth, K. Abbas, F. Simonelli, J. Kozempel, I. Cydzik, G. Cotogno, A. Bulgheroni, D. Gilliland, J. Ponti, F. Franchini, P. Marmorato, H. Stamm, W. Kreyling, A. Wenk, M. Semmler-Behnke, S. Buono, L. Maciocco, and N. Burgio. 2011. Radiolabelling of engineered nanoparticles for in vitro and in vivo tracing applications using cyclotron accelerators. Arch. Toxicol. 85(7):751-773.

Goble, C.A., J. Bhagat, S. Aleksejevs, D. Cruickshank, D. Michaelides, D. Newman, M. Borkum, S. Bechhofer, M. Roos, P. Li, and D. De Roure. 2010. myExperiment: A repository and social network for the sharing of bioinformatics workflows. Nucl. Acids Res. 38:W677-W682.

Gordon, N., and U. Sagman. 2003. Nanomedicine Taxonomy Briefing Paper. Canadian NanoBusiness Alliance, Toronto, Ontario. February 2003 [online]. Available: http://www.nanowerk.com/nanotechnology/reports/reportpdf/report31.pdf [accessed June 1, 2011].

Gottschalk, F., and B. Nowack. 2011. The release of engineered nanomaterials to the environment. J. Environ. Monit. 13(5):1145-1155.

Gottschalk, F., T. Sonderer, R.W. Scholz, and B. Nowack. 2009. Modeled environmental concentrations of engineered nanomaterials (TiO_2, ZnO, Ag, CNT, fullerenes) for different regions. Environ. Sci. Technol. 43(24):9216-9222.

Hackley, V.A., M. Fritts, J.F. Kelly, A.K. Patri, and A.F. Rawle. 2009. Enabling standards for nanomaterial characterization. Pp. 24-29 in INFOSM Informative Bulletin of the Interamerican Metrology System. August 13, 2009.

Hamadeh, H.K., P.R. Bushel, S. Jayadev, O. DiSorbo, L. Bennett, L. Li, R. Tennant, R. Stoll, J.C. Barrett, R.S. Paules, K. Blanchard, and C.A. Afshari. 2002. Prediction of compound signature using high density gene expression profiling. Toxicol. Sci. 67(2):232-240.

Hassellöv, M., J.W. Readman, J.F. Ranville, and K. Tiede. 2008. Nanoparticle analysis and characterization methodologies in environmental risk assessment of engineered nanoparticles. Ecotoxicology 17(5):344-361.

Hong, H., Y. Zhang, J. Sun, and W. Cai. 2009. Molecular imaging and therapy of cancer with radiolabeled nanoparticles. Nano Today 4(5):399-413.

IANH (International Alliance for NanoEHS Harmonization). 2011. International Alliance for NanoEHS Harmonization [online]. Available: http://www.nanoehsalliance.org/sections/Home [accessed May 13, 2011].

InterNano. 2011. InterNano. Resources for Manufacturing. Nanoinformatics 2020 Roadmap [online]. Available: http://eprints.internano.org/607/ [accessed Nov. 9, 2011].

Jarvie, H.P., H. Al-Obaidi, S.M. King, M.J. Bowes, M.J. Lawrence, A.F. Drake, M.A. Green, and P.J. Dobson. 2009. Fate of silica nanoparticles in simulated primary wastewater treatment. Environ. Sci. Technol. 43(22):8622-8628.

Jeong, C.H., P.K. Hopke, D. Chalupa, and M. Utell. 2004. Characteristics of nucleation and growth events of ultrafine particles measured in Rochester, NY. Environ. Sci. Technol. 38(7):1933-1940.

Jillavenkatesa, A., S.J. Dapkunas, and L.S.H. Lum. 2001. Particle Size Characterization. Special Publication 960-1. U.S. Department of Commerce, National Institute of Standards and Technology, Gaithersburg, MD [online]. Available: http://www.nist.gov/public_affairs/practiceguides/SP960-1.pdf [accessed Nov. 10, 2011].

Johnston, J.M., M. Lowry, S. Beaulieu, and E. Bowles. 2010. State-of-the-Science Report on Predictive Models and Modeling Approaches for Characterizing and Evaluating Exposure to Nanomaterials. EPA/600/R-10/129. U.S. Environmental Protection Agency, Washington, DC [online]. Available: http://www.epa.gov/athens/publications/reports/Johnston_EPA600R10129_State_of_Science_Predictive_Models.pdf [accessed Nov. 23, 2011].

Klaper, R., and M.A. Thomas. 2004. At the crossroads of genomics and ecology: The promise of a canary on a chip. BioScience 54(5):403-412.

Klemm, J., N. Baker, D. Thomas, S. Harper, M.D. Hoover, M. Fritts, R. Cachau, S. Gaheen, S. Pan, G. Stafford, and D. Paik. 2010. nano-TAB: A Standard File Format for Data Submission and Exchange on Nanomaterials and Characterizations. Nanoinformatics

Conference November 3-5, 2010, Arlington, VA [online]. Available: http://nanotechinformatics.org/posters [accessed Apr. 22, 2011].

Kozaki, K., Y. Kitamura, and R. Mizoguchi. 2011. Systematization of Nanotechnology Knowledge through Ontology Engineering: A Trial Development of Idea Creation Support System for Materials Design based on Functional Ontology. Poster Notes of the Second International Semantic Web Conference (ISWC 2003), October 20-23, 2003, Sanibel Island, FL [online]. Available: http://www.ei.sanken.osaka-u.ac.jp/pub/kozaki/iswc2003pos_kozaki.pdf [accessed Apr. 22, 2011].

Kreuter, J. 1991. Nanoparticle-based drug delivery systems. J. Control. Rel. 16(1-2):169-176.

Leeuw, T.K., R.M. Reith, R.A. Simonette, M.F. Harden, P. Cherukuri, D.A. Tsyboulski, K.M. Beckingham, and R.B. Weisman. 2007. Single-walled carbon nanotubes in the intact organism: Near-IR imaging and biocompatibility studies in Drosophila. Nano Lett. 7(9):2650-2654.

Lynch, I., T. Cedervall, M. Lundqvist, C. Cabaleiro-Lago, S. Linse, and K.A. Dawson. 2007. The nanoparticle - protein complex as a biological entity; A complex fluids and surface science challenge for the 21st century. Adv. Colloid Interface Sci. 134-35:167-174.

Metz, K.M., A.N. Mangham, M.J. Bierman, S. Jin, R.J. Hamers, and J.A. Pedersen. 2009. Engineered nanomaterial transformation under oxidative environmental conditions: Development of an in vitro biomimetic assay. Environ. Sci. Technol. 43(5):1598-1604.

MINChar Initiative. 2009. Characterization Matters: Supporting Appropriate Material Characterization in Nanotoxicology Studies [online]. Available: http://characterizationmatters.org/ [accessed May 12, 2011].

Monopoli, M.P., D. Walczyk, A. Campbell, G. Elia, I. Lynch, F.B. Bombelli, and K.A. Dawson. 2011. Physical-chemical aspects of protein corona: Relevance to in vitro and in vivo biological impacts of nanoparticles. J. Am. Chem. Soc. 133(8):2525-2534.

Morawska, L., C. He, G. Johnson, H. Guo, E. Uhde, and G. Ayoko. 2009. Ultrafine particles in indoor air of a school: Possible role of secondary organic aerosols. Environ. Sci. Technol. 43(24):9103-9109.

Murashov, V., and J. Howard, eds. 2011. Pp. 196-197 in Nanotechnology Standards. Nanostructure Science and Technology Series. New York: Springer.

Nel, A.E., L. Madler, D. Velegol, T. Xia, E.M. Hoek, P. Somasundaran, F. Klaessig, V. Castranova, and M. Thompson. 2009. Understanding biophysicochemical interactions at the nano-bio interface. Nat. Mater. 8(7):543-557.

NIST/NCL (National Institute of Standards and Technology and Nanotechnology Characterization Laboratory). 2010. Measuring the Size of Nanoparticles Using Transmission Electron Microscopy (TEM). NIST-NCL Joint Assay Protocol, PCC-7. Version 1.1. U.S. Department of Commerce, National Institute of Standards and Technology, Gaithersburg, MD, and National Cancer Institute, Nanotechnology Characterization Laboratory, Frederick, MD [online]. Available: http://ncl.cancer.gov/NCL_Method_PCC-7.pdf [accessed May 12, 2011].

NRC (National Research Council). 2007. Toxicity Testing in the 21st Century: A Vision and a Strategy. Washington, DC: The National Academies Press.

NRC (National Research Council). 2009. Science and Decisions: Advancing Risk Assessment. Washington, DC: The National Academies Press.

Oberdörster, G., A. Maynard, K. Donaldson, V. Castranova, J. Fitzpatrick, K. Ausman, J. Carter, B. Karn, W. Kreyling, D. Lai, S. Olin, N. Monteiro-Riviere, D. Warheit, and H. Yang. 2005. Principles for characterizing the potential human health effects from exposure to nanomaterials: Elements of a screening strategy. ILSI research foundation/Risk Science Institute Nanomaterial Toxicity Screening Working Group. Part. Fibre Toxicol. 2:8.

OECD (Organisation for Economic Co-operation and Development). 2010. Guidance Manual for the Testing of Manufactured Nanomaterials OECD's Sponsorship Programme; First Revision. ENV/JM/MONO(2009)20/REV. Organization for Economic Co-operation and Development. June 2, 2010 [online]. Available: http://www.oecd.org/LongAbstract/0,3425,en_2649_34365_45409513_1_1_1_1,00.html [accessed Apr. 25, 2011].

Oostingh, G.J., E. Casals, P. Italiani, R. Colognato, R. Stritzinger, J. Ponti, T. Pfaller, Y. Kohl, D. Ooms, F. Favilli, H. Leppens, D. Lucchesi, F. Rossi, I. Nelissen, H. Thielecke, V.F. Puntes, A. Duschl, and D. Boraschi. 2011. Problems and challenges in the development and validation of human cell-based assays to determine nanoparticle-induced immunomodulatory effects. Part. Fibre Toxicol. 8(1):8.

OSI (OSI e-Infrastrustructure Working Group). 2007. Developing the UK's e-Infrastructure for Science and Innovation, Report of the OSI e-Infrastructure Working Group, National e-Science Center. January 18, 2007 [online]. Available: http://immagic.com/eLibrary/ARCHIVES/GENERAL/NESC_UK/N070118O.pdf [accessed May 25, 2011].

Ostraat, M. 2011. The Nanomaterial Registry. Presentation at the Society for Toxicology Annual Meeting, March 6-10, 2011, Washington, DC [online]. Available: http://www.toxicology.org/isot/ss/nano/docs/Ostraat_guest_presentation.pdf [access May 12, 2011].

Petersen, E.J., Q.G. Huang, and W.J. Weber, Jr. 2008. Bioaccumulation of radio-labeled carbon nanotubes by Eisenia foetida. Environ. Sci. Technol. 42(8):3090-3095.

Petosa, A.R., D.P. Jaisi, I.R. Quevedo, M. Elimelech, and N. Tufenkji. 2010. Aggregation and deposition of engineered nanomaterials in aquatic environments: Role of physicochemical interactions. Environ. Sci. Technol. 44(17):6532-6549.

Phenrat, T., N. Saleh, K. Sirk, H.J. Kim, R.D. Tilton, and G.V. Lowry. 2008. Stabilization of aqueous nanoscale zerovalent iron dispersions by anionic polyelectrolytes: Adsorbed anionic polyelectrolyte layer properties and their effect on aggregation and sedimentation. J. Nanopart. Res. 10(5):795-814.

Phenrat, T., T.C. Long, G.V. Lowry, and B. Veronesi. 2009. Partial oxidation ("aging") and surface modification decrease the toxicity of nanosized zerovalent iron. Environ. Sci. Technol. 43(1):195-200.

Richman, E.K., and J.E. Hutchison. 2009. The nanomaterial characterization bottleneck. ACS Nano. 3(9):2441-2446.

Saleh, N., H.J. Kim, T. Phenrat, K. Matyjaszewski, R.D. Tilton, and G.V. Lowry. 2008. Ionic strength and composition affect the mobility of surface-modified Fe^0 nanoparticles in water-saturated sand columns. Environ. Sci. Technol. 42(9):3349-3355.

Schierz, P.A., A.N. Parks, and P.L. Ferguson. 2010. Characterization and Analysis of Single-walled Carbon Nanotubes in Complex Matrices by Asymmetric Flow FFF Coupled with NIRF Spectroscopy. Presentation at the 5th Annual Conference on Environmental Effects of Nanoparticles and Nanomaterials, August 2010, Clemson, SC.

Schmieder, P.K., G. Ankley, O. Mekenyan, J.D. Walker, and S. Bradbury. 2003. Quantitative structure-activity relationship models for prediction of estrogen receptor binding affinity of structurally diverse chemicals. Environ. Toxicol. Chem. 22(8):1844-1854.

Smith, J.N., K.C. Barsanti, H.R. Friedli, M. Ehn, M. Kulmala, D.R. Collins, J.H. Scheckman, J. Williams, and P.H. McMurry. 2010. Observations of aminium salts in atmospheric nanoparticles and possible climatic implications. Proc. Natl. Acad. Sci. U.S.A. 107(15):6634-6639.

Tan, W., R. Madduri, A. Nenadic, S. Soiland-Reyes, D. Sulakhe, I. Foster, and C. Goble. 2010. CaGrid Workflow Toolkit: A taverna based workflow tool for cancer grid. BMC Bionformatics 11:542.

Teeguarden, J.G., P.M. Hinderliter, G. Orr, B.D. Thrall, and J.G. Pounds. 2007. Particokinetics in vitro: Dosimetry considerations for in vitro nanoparticle toxicity assessments. Toxicol. Sci. 95(2):300-312.

Thomas, D.G., R.V. Pappu, and N.A. Baker. 2011. NanoParticle Ontology for cancer nanotechnology research. J. Biomed. Inform. 44(1):59-74.

Tiede, K., M. Hassellöv, E. Breitbarth, Q. Chaudhry, and A.B. Boxall. 2009. Considerations for environmental fate and ecotoxicity testing to support environmental risk assessments for engineered nanoparticles. J. Chromatogr. A. 1216(3):503-509.

Vietti-Cook, A.L. 1999. Staff Requirements - SECY-98-144 - White Paper on Risk-Informed and Performance-Based Regulation. Memorandum to William D. Travers, Executive Director for Operations, from Annette L. Vietti-Cook, Secretary, U.S. Nuclear Regulatory Commission. March 1, 1999 [online]. Available: http://pbadupws.nrc.gov/docs/ML0037/ML003753601.pdf [accessed Nov. 28, 2011].

von der Kammer, F., P.L. Ferguson, P.A. Holden, A. Masion, K.R. Rogers, S.J. Klaine, A.A. Koelmans, N. Horne, and J.M. Unrine. 2012. Analysis of engineered nanomaterials in complex matrices (environment & biota): General considerations and conceptual case studies. Environ. Toxicol. Chem. 31(1):32-49.

W3C Semantic Web. 2011. W3C Semantic Web Activity [online]. Available: http://www.w3.org/2001/sw/ [accessed May 16, 2011].

Walczyk, D., F.B. Bombelli, M.P. Monopoli, I. Lynch, and K.A. Dawson. 2010. What the cell "sees" in bionanoscience. J. Am. Chem. Soc. 132(16):5761-5768.

Wiesner, M.R., G.V. Lowry, P. Alvarez, D. Dionysiou, and P. Biswas. 2006. Assessing the risks of manufactured nanomaterials. Environ. Sci. Technol. 40(14):4336-4345.

Wiesner, M.R., G.V. Lowry, K.L. Jones, M.F. Hochella, Jr., R.T. Di Giulio, E. Casman, and E.S Bernhardt. 2009. Decreasing uncertainties in assessing environmental exposure, risk, and ecological implications of nanomaterials. Environ. Sci. Technol. 43(17):6458-6462.

Zhao, J., F.L. Eisele, M. Titcombe, C.G. Kuang, and P.H. McMurry. 2010. Chemical ionization mass spectrometric measurements of atmospheric neutral clusters using the cluster-CIMS. J. Geophys. Res. 115:D08205.

5

Research Priorities and Resource Needs

OVERVIEW

Chapter 2 developed a conceptual framework (Figure 2-1) for considering environmental, health, and safety (EHS) risks related to nanomaterials and to help prioritize activities within a strategic research plan. Three overarching principles were developed to guide the development of strategic research and to identify ENMs requiring particular attention; emergent risk, plausibility and severity. In addition, specific criteria were established as a basis for assigning research priorities:

- Research that advances knowledge of both exposure and hazard wherever possible.
- Research that leads to the production of risk information needed to inform decision-making on nanomaterials in the market place.
- Research efforts to address short-term needs that serve as a foundation for moving beyond case-by-case evaluations of nanomaterials and allow longer-term forecasting of risks posed by newer materials expected to enter commerce.
- Research that promotes the development of critical supporting tools, such as measurement methods, limitations of which hinder the conduct of research in processes that control hazards and exposure.
- Research on ecosystem-level effects that addresses exposure or hazard scenarios that are underrepresented in the current portfolio of nanotechnology-related EHS research; for example, impacts on ecosystem processes and on organisms representing different phyla and environments.

Chapter 3 reviewed what is known about the EHS aspects of nanomaterials in the context of the conceptual framework and identified critical research questions that remain unanswered, focusing on processes most likely to affect exposure and hazards related to engineered nanomaterials (ENMs) (circle in Figure 2-1). Chapter 4 addressed tools needed for characterizing how the properties of ENMs affect their interactions with humans and the environment (bottom of Figure 2-1).

In approaching the charge to develop research priorities, the committee applied the framework developed in Chapter 2 to the research and development needs identified in Chapters 3 and 4, and in doing so, identified four broad, cross-cutting research priority categories. These mirror the larger elements of the conceptual framework described in Chapter 2 and map directly to the critical research needs identified in Chapters 3 and 4. At the chapter's end, the committee discusses the resources needed to implement a strategic research plan within the context of these priority categories. The research categories are

- Adaptive research and knowledge infrastructure for accelerating research progress and providing rapid feedback to advance research.
- Characterizing and quantifying the origins of nanomaterial releases.
- Processes affecting *both* potential hazard and exposure.
- Nanomaterial interactions in complex systems ranging from subcellular systems to ecosystems.

Given the diversity of nanomaterials and the breadth of their potential applications, the committee considered that a prescriptive approach to addressing the EHS aspects of nanomaterials would be short-sighted and would probably fail to anticipate the rapid evolution of this field and its potential impacts. Rather, in selecting the four broad categories, the committee envisioned a risk-based system that is iteratively informed and shaped by the outcomes of research and new findings.

Thus, its approach addresses one goal in particular as described in Chapter 1: to generate scientific evidence that provides approaches to environmental and human health protection even as our knowledge of ENMs is expanding and the research strategy is evolving. Furthermore, as the research strategy is evolving, an adaptive and integrated knowledge infrastructure will be developed to identify and enable prediction of risks posed by nanomaterials with sufficient certainty to enable informed decisions on how the risks should be managed or mitigated. The knowledge infrastructure also will provide evidence that helps to identify and evaluate the merits of various risk-management options, including measures to reduce inherent hazard or exposures to nanomaterials.

The committee proposes a strategy to address the EHS aspects of nanomaterials that sets priorities for research efforts that bridge complex and model systems, exposure and hazard, and immediate and long-term concerns, thus reflecting the need for systems integration described in Chapter 3. The strategy also favors the development of supporting measurement and modeling tools to advance the study of nanomaterial interactions among the risk-assessment domains of exposure, hazard, and risk characterization. Such broad, overarching priorities were deemed important by the committee, given the relevance of nanomaterials to numerous scientific and technical disciplines—including electronics, energy, and medicine—and the array of exposure and hazard scenarios. The committee also identified commonalities in the research tools (for example, measurement methods and data infrastructure) throughout multiple levels of organization (for example cellular, organismal, or ecosystem) that can be capitalized on in addressing research priorities.

In the sections below, the committee expands on each of the four research priority categories and describes their relationship to the data gaps and key research questions in Chapters 3 and 4. The four categories are of equal priority and interconnected; their ordering does not imply a priority, and some research components are common to all four priority categories. In some cases, the committee describes components of the categories that need to be addressed in the short term and components that will evolve. A short-term timeframe is considered to be within 5 years. The priorities are activities that can be readily organized, resourced, and accomplished with available knowledge and tools. They need to be accomplished because they are fundamental to informing or enabling other activities. Furthermore, many topics on which research is expected to be initiated in the short term will continue to be addressed in the longer term as new tools and approaches are developed; this emphasizes the iterative nature of the research strategy.

Because of the iterative and sequential nature of the research process, the committee demarcated short-term and long-term research only when there was an evident distinction in timing. The committee describes the logical sequence of the research within each of the priority categories with recognition that timing will depend on the knowledge gained from previous research efforts.

ADAPTIVE RESEARCH AND KNOWLEDGE INFRASTRUCTURE FOR ACCELERATING RESEARCH PROGRESS AND PROVIDING RAPID FEEDBACK TO ADVANCE RESEARCH

An adaptive knowledge infrastructure is essential for supporting and providing rapid feedback on integrative research. Broadly, the infrastructure must support the generation of inputs (materials, methods, and end points), the development of relationships and models based on data-sharing and validation of the models, and the development of hypotheses and predictions from the models. The infrastructure encompasses tools, including materials, characterization methods, models, and informatics.

The infrastructure should also

- Identify emerging data gaps and highlight those that need to be addressed.
- Provide rapid feedback to inform research and design of new materials with reduced hazards or exposure potential.
- Be accessible to the public and to scientists.

The outcomes needed from the research and knowledge infrastructure include making characterized nanomaterials widely available, refining analytic methods continuously to define the structures of the materials throughout their life span, defining methods and protocols to assess effects, and increasing the rate of generation and the quality of the data and models available. Stakeholders should be engaged in developing best practices, in sharing information, and in collaborating in developing methods and models. Informatics should be fostered through the joining ("federating") of existing databases, the encouraging and sustaining of curation and annotation of the data, and the assigning of credit to those who share datasets and models. Joined knowledge bases need to be interoperable and provide for mapping or translation of related ontologies (descriptions of the concepts and relationships among a set of agents or elements; Gruber 2011) to allow for searching similar concepts to identify appropriate data.

Because the knowledge infrastructure will integrate the research agenda, it comprises activities that connect the other research categories. Each activity described below is an integral part of the infrastructure and has both short-term and long-term components. The relative emphasis placed on each activity will change as the foundational components of the infrastructure are established; however, to ensure coordination of the infrastructure, some emphasis on each activity is needed from the outset. The activities are binned into three areas, in descending order of their importance in the short term.

Short-term priority requiring immediate emphasis followed by a sustained effort:

- Produce and make available material libraries (characterized nanomaterials in commerce, reference materials, and standard materials) that have the structural definition and systematic variation needed for advancing key research (see Chapter 4).

Building these capabilities in the short term and ramping them up to a sustained effort in the longer term:

- Develop and validate the analytic tools and methods needed to relate nanomaterial properties to system responses, including methods for detecting, characterizing, and tracking nanomaterials in relevant media and for monitoring transformations (including surface modifications) in complex media and on the timescale of experiments. A multi-tiered approach will be needed to develop methods so that the fate of ENMs in all relevant media can be understood.
- Refine and validate methods needed to characterize and quantify the effects of ENMs in experimental systems, considering the identity and dose of a nanomaterial at the target or in the system. Develop and validate methods, including high-throughput screening, to examine the sensitivity of effects to structural motifs and descriptors.

• Create and support mechanisms for data-sharing to advance research and to generate understanding of relationships among data using models. Data-sharing systems should be collaborative and nimble; should engage the broad array of stakeholders in producing materials, instruments, and models; and should provide mechanisms for sharing raw data and results (negative and positive). The findings should support continuing research and the design of safer nanomaterials.

Longer-term efforts that require consideration and coordination in the short term to ensure that experimental, modeling, and informatics efforts contribute to a coordinated, functional infrastructure

• Advance and validate models for nanomaterials, nanomaterial transformations, and target systems that test specific and systemwide effects of ENMs.
• Encourage collaboration between experimentalists and computer modelers and develop descriptors to compare materials, models, and model results.
• Establish and evolve an informatics framework that begins by federating and supporting existing data repositories and connecting them through shared or translatable ontologies.

CHARACTERIZING AND QUANTIFYING THE ORIGINS OF NANOMATERIAL RELEASES

Characterizing the quantity and nature of nanomaterials to which human populations and ecosystems are exposed is critical for evaluating the EHS risks posed by nanomaterials. Exposure to a nanomaterial in any setting (for example, in the workplace, in domestic use, or in the environment) is a result of conditions associated with the initial state of the nanomaterial (for example, as individual particle vs embedded in a synthetic or biologic matrix), the potential pathways of exposure, and the influences of environmental conditions (such as temperature, sunlight, and flow of air or water) on the nanomaterial released. An understanding of release scenarios across the life cycle and value chain of the material is needed, including in production, in use, and at the end of life. For example, as discussed in Chapter 3, characterizing the nature and relevance of exposure requires information on the sources of exposure (including nanomaterials and contaminants), the nature of nanomaterials in products, the condition of the material released, the release points where nanomaterials may enter the environment (for example, in transportation, waste-handling, product recycling, and disposal), and the environmental conditions in which releases occur. Just as complex systems in nature (such as organisms and ecosystems) present challenges for detection and characterization of nanomaterials across the value chain, so do human activities. Moreover, socioeconomic drivers, proprietary considerations, and complex networks of products and information dissemination support the need for an integrative-research structure. Industry involvement in this research to understand trends in manufacturing and "horizon materials" will probably be a key input to advancing this priority category.

In summary, research activities in this category would

- Identify and characterize what is being released and who is exposed:
  - Identify critical release points and quantities along the life cycle, including the waste streams, and characterize materials being released.
  - Identify human and ecosystem populations exposed.
  - Characterize releases and exposures of workers and consumers of high-production nanomaterials currently in the market and in use down the value chain (for example, carbon nanotubes).
- Define the range of materials to
  - Develop inventories of current and expected production of nanomaterials[1].
  - Develop inventories of current and expected use of nanomaterials and value-chain transfers.
- Measure the quantity and nature of released materials in associated receptor environments to
  - Quantify exposures.
  - Model nanomaterial releases along the life cycle.

Short-term activities address materials in commerce and in the environments that nanomaterials will enter along the value chain and lifecycle. Exposure routes and transformations can be determined on the basis of uses of commercially available nanomaterials and their properties and formats (for example, their presence in emulsions or polymer matrices). Longer-term activities address trends in nanomaterial markets and in nanomaterial development; new materials and new markets create greater uncertainty regarding the nature of the nanomaterials, production quantities, uses, and routes of exposure. Longer-term activities also will consider other life-cycle impacts of nanomaterial production, such as material and energy use and the wastes produced.

Short-Term Activities

- Inventories of current and near-term production of nanomaterials.

---

[1]The committee identified the need for better understanding of ENMs that are in or about to enter commerce, and the development of inventories of nanomaterial production and use to that end is warranted. To the extent that reporting by industry would contribute to such an effort, the entities and nanomaterials subject to reporting would need to be defined with some precision. Those issues were addressed in establishing the inventory reporting systems developed for conventional chemicals (for example, EPA's Toxic Substances Control Act Chemical Data Reporting Program [EPA 2011]).

- Inventories of intended use of nanomaterials and value-chain transfers.
- Identification of critical release points along the value chain.
- Identification of populations or systems exposed.
- Characteristics of released materials and associated receptor environments.
- Modeling of nanomaterial releases along the value chain.

Middle-Term to Long-Term Activities

- Development of models to anticipate trends in production and use of nanomaterials and characteristics of future releases.
- Development of a more sophisticated understanding of the release of and exposure to more complex "new-generation" nanomaterials, such as those involving composite materials, active-bioactive nanomaterials, and composites of biologic materials and nanomaterials.
- Life-cycle analysis of nanomaterial production, use, and disposal with an accounting of energy and material inputs and the wastes produced.

PROCESSES AFFECTING BOTH EXPOSURE AND HAZARD

Because nanoscale properties have a profound influence on biologic, physical, and chemical processes that control nanomaterial releases, transformations, and effects in various levels of biologic organization, from organisms to ecosystems, it is advantageous to assess exposure to and hazards of nanomaterials together. The conceptual framework for the EHS research strategy (Figure 2-1) emphasized the investigation of processes common to exposure and hazard as a basis for advancing risk assessment in a predictive and generalizable fashion, thereby laying a foundation for informing decision-making on current and future nanomaterials. Research in this category is focused on the nanoscale where advances in information technologies, the life sciences, physical chemistry, materials science, and other disciplines converge. The common ground for interdisciplinary research at the nanoscale should be exploited through development of fundamental knowledge that advances our understanding of both exposures and hazards.

As discussed in Chapter 2, an approach that simultaneously addresses exposure and hazard enables decision-making in the short to long term related to comparing risks among materials and providing criteria for establishing priorities for research on nanomaterials that are currently on the market, to providing feedback on research needs and priorities, and to providing information needed to reduce the risks posed by nanomaterials that are on the market or are under development.

One example of the importance of understanding both hazards and exposures is the role of nanoparticle-macromolecular interactions in regulating and modifying nanoparticle behavior at scales ranging from genes to ecosystems. Such interactions result from nanomaterial properties that originate during their creation with adsorbed coatings (for example, materials designed to stabilize particles against aggregation, enhance their association with targeted cells in drug delivery, improve their dispersion in emulsions, or contain their reactivity, as in sunscreens). Physiologic and ecologic environments may modify the material surfaces, for example, through adsorption of proteins in blood or naturally occurring polyelectrolytes in surface waters. Studying nanoparticle-macromolecular interactions as a unified topic rather than separately as an investigation of hazard or exposure can lead to a better understanding of the processes controlling nanoparticle mobility, environmental partitioning, biodistribution, bioaccumulation, protein folding, and changes in the conformation of RNA or DNA with associated changes in gene expression. The cross-disciplinary interactions among physical chemists, toxicologists, geochemists, molecular biologists, and other scientists are critical for the study of processes at the nanoscale, such as particle aggregation, deposition on surfaces, reactivity, persistence, and biokinetics.

Topics in this research category include the effects of particle surface modification on aggregation and nanoparticle bioavailability, reactivity, and toxicity potential; processes that affect nanomaterial transport across biologic or synthetic membranes; and the development of structure-activity relationships of nanomaterials with their transport, fate, and effects. For example, surface modification of zero-valent iron [Fe(0)] nanoparticles with polyelectrolyte or natural organic matter reduces their aggregation and deposition and enhances their transport in soils (Saleh et al. 2008; Johnson et al. 2009). Decreasing aggregation and increasing mobility in the environment increases the potential for exposure to those materials. However, the same surface modifications decrease the toxicity of the materials to bacteria (Li et al. 2010) and to neurons and central nervous system microglia cells (Phenrat et al. 2009a) compared with unmodified particles. Despite the decrease in toxicity based on the end points studied, coated particles were able to enter the cells' nuclei, and uncoated (aggregated) Fe(0)/Fe_3O_4 core-shell nanoparticles were not (Phenrat et al. 2009a). Similarly, the oxidation of Fe(0)/Fe_3O_4 core-shell to Fe-oxide increases the mobility of the nanoparticles in the environment but dramatically decreases their toxicity to multiple receptors (Auffan et al. 2008; Phenrat et al. 2009b). Thus, such processes as surface modification, aggregation, and oxidation can in this case increase the potential for exposure but decrease hazard. Ultimately, understanding how those fundamental processes affect risk will inform risk-management decisions about nanomaterial compositions, modifications, and application scenarios.

Research in this category is necessarily broad. Some examples of activities are

- Development of instrumentation and standard methods for characterizing releases and exposures to ENMs in relevant biologic and environmental media, including ENM size, surface character, and composition dynamically (from seconds to hours) and single particles.

- Studies that specify and characterize trends in how ENM transformations influence the biologic effects of ENMs on organisms and ecosystems.
- Cross-cutting research that systematically describes ENM transformations that occur in organisms or as a result of biologic processes.
- Examination of how native ENM structures influence the dynamic ENM structures that develop in environmental and biologic settings.
- Research that provides a generalized and quantitative understanding of ENM transformations through the development of predictive models.
- Further development of a knowledge infrastructure that can describe and allow for the diversity and dynamics of ENM structure in relevant biologic and environmental media.

Instrumentation to measure ENM properties in various matrices is needed to relate their properties to the potential for exposure and effects and to determine the types and extent of ENM transformations in environmental and biologic systems. Because unique properties of a subset of ENMs may cause them to behave differently from other nanomaterials, methods for characterizing single-particle features of the ENMs are needed. All the activities discussed are considered to have high priority, but initial investigations should be weighted toward the development of characterization methods, including for single particles, so that the methods will be available for addressing issues related to ENM structure and activity and to transformations in environmental and biologic media.

NANOMATERIAL INTERACTIONS IN COMPLEX SYSTEMS RANGING FROM SUBCELLULAR SYSTEMS TO ECOSYSTEMS

EHS research on ENMs is unified by the need to understand their interactions with complex systems, whether subcellular components, single cells, organisms, or ecosystems. Each of those systems has a level of complexity with many embedded, interrelated processes that may interact synergistically, antagonistically, and often unpredictably in response to the introduction of nanomaterials. The scientific community has recognized the need for system-level approaches to understand the potential for ENM effects on human health and the environment. That recognition encompasses the notion of indirect consequences of direct interactions. For example, effects on environmental geochemistry can affect the viability of key components of the ecosystem food web that ultimately may affect ecosystem integrity. In addressing responses to ENMs, whether at the cellular level or the ecosystem level, there are common challenges, including a need to re-examine "default" assumptions and scenarios. The challenges include classic examples from ecosystem science and toxicology, such as extrapolation from high-dose effects to low-dose effects, temporality of response, generalizability from one animal model (or ecosystem) to another, and variability within populations (or habitats). Transformations that occur in physiologic or environmental systems, such as weathering (ecosystems) and metabolism (organisms and ecosystems), and interactions with macromolecules (for example, blood serum proteins, DNA, humic materials, and polysaccharides) introduce complexity that must be considered in performing research.

Traditional end points and associated metrics (for example, LD_{50}) may not capture more subtle effects that occur in the context of development, reproduction, repair, adaptation, and behavior, given population characteristics and individual variability. Indeed, there are probably ecosystem effects that cannot be predicted from single-organism toxicity tests; this is analogous to the failure to predict human toxicologic effects from single responses, such as inflammation, particularly when such effects are measured at a single time or at an unrealistically high dose. In this priority category, research includes efforts to relate in vitro to in vivo observations, predicting such system-level effects as nutrient cycling, and at the organism level, assessing effects on the endocrine or developmental systems.

This category also encompasses the need to develop a more rigorous, conceptual, and complex model of effects of exposure and potential effects along the ecologic food chain concomitantly with corresponding development of instrumentation and protocols for nanomaterial measurement, and detection and development of assays for isolating ENM effects and reactivity in complex media. Lessons learned from complex systems, such as the relevance of characteristics and interferences for predicting physiologic and environmental effects of nanoparticles, must be systematically fed back into experimental-design efforts, such as high-throughput screening. Research in this category therefore must be supported by the development of informatics tools and the knowledge infrastructure. At some point, models and physicochemical data must be validated with experimental data.

To identify the most productive subjects for research, a matrix approach that takes advantage of the development of tools for nanomaterial characterization and systems effects should be considered. The matrix elements would include in one dimension screening tools for determining nanomaterial effects, ranging from subcellular to ecosystems, and in the other dimension a set or library of standard nanomaterials characterized by specific properties, such as composition, surface charge, size, and shape. From those complex, system level models or screening approaches would be derived nanomaterial and system characteristics. These characteristics would indicate directions for exploration of mechanisms of effects to develop predictive methods for assessing effects that depend on the combinations of nanomaterial and system properties. Nothing in this approach would impede research that is successfully exploring known interactions of nanomaterials in complex systems from subcellular systems to ecosystems.

Short-Term Activities

- Refinement of a set of screening tools (high-throughput and high-content) that reflect important characteristics and toxicity pathways of the complex systems described above.
- Adaptation of system-level tools (for example, individual species tests, microcosms, and organ-system models) to support in vitro to in vivo correlations for nanomaterials, including exposure route, dose, and mechanisms of effects.
- Development of approaches for comparing standardized reference nanomaterials with a variety of traditional toxic substances to understand such issues as bioavailability, metabolism, and relative potency.
- On the basis of results of the above activities, identification of reference materials that can be used as positive or negative controls for a variety of high-throughput and high-content test systems.
- Development of tools and methods for estimating exposure and doses in complex systems, including approaches to address portal of entry, kinetics of toxicity, and mechanisms of effects.
- Identification of potentially exposed, susceptible human and ecosystem populations with a focus on end points identified through previous studies and development of surveillance tools for detecting and characterizing effects of nanomaterials in these populations and ecosystems.

Long-Term Activities

Long-term priorities will depend on the successful completion of some of or all the short-term activities detailed above. An example of a longer-term priority is research on complex human health effects, in which it is possible that low-dose inhalation exposures to nanomaterials over a long term could have pathologic sequelae from sustained (low-level) pulmonary inflammation to pulmonary fibrosis. In that example, it would be critical to identify nanomaterial-specific dose-related and time-course-related mechanisms of action in vivo to understand toxicokinetic and toxicodynamic characteristics of nanomaterials in order to facilitate development of high-throughput testing and high-content screening with in vitro and in silico methods. The development of those approaches could be expedited by inclusion of standard reference nanomaterials in experimental testing designs, but such approaches must await the development of short-term data to support testing strategies to advance system-level understanding of the potential EHS effects of nanomaterials.

Research in the four broad priority categories will address the goals of the research agenda articulated in Chapter 1 by generating scientific evidence that

- Provides for approaches to environmental and human health protection even as our knowledge of ENMs is expanding and the research strategy is evolving.

- Identifies and predicts risks posed by nanomaterials with sufficient certainty to enable informed decisions on how the risks should be prevented, managed, or mitigated.
- Identifies and evaluates the relative merits of various risk-management options, including measures to reduce the inherent hazard or exposure potential of nanomaterials.

RESOURCES FOR ADDRESSING RESEARCH PRIORITIES

In addition to identifying the four research priority categories, the committee considered the resources needed to address its recommendations, consistent with its charge. In making these recommendations, the committee recognizes the current funding situation and the overall inadequacy of the funding available. Given this constraint, the committee provided pragmatic and general recommendations for funding. While it mentions specific amounts, the guidance should not be construed as reflecting the priority that should be given to a particular category, as we considered all categories of equal priority. There is, however, a sequencing to these categories that is reflected in the resource recommendations.

In making its resource recommendations, the committee examined past funding levels, but did not conduct a more formal analysis of funding. In its second report, the committee will revisit its funding recommendations, based on further analysis and changes in the funding context.

Overall, there has been concern as to whether the level of federal funding devoted to EHS research related to nanomaterials is sufficient (Denison 2005a; GAO 2008; Maynard 2008; Sargent 2011). That concern was echoed in the NRC (2009) review of the federal strategy (NEHI 2008). However, as in all areas of research, placing a dollar investment value on needed research is a necessarily complex and qualitative approach, made more difficult by uncertainties in potential adverse impacts in the absence of research investment, in the potential for reducing adverse impacts with findings from research investments, and in the application of research findings in associated fields (for example, the use of information resulting from nanoscale-medical applications to address the EHS impacts of other ENMs).

In 2006, The Project on Emerging Nanotechnologies conducted an assessment of nanotechnology-related EHS research gaps and recommended a minimum EHS R&D investment of $100 million over the following 2 years to address highly targeted risk-based research (Maynard 2006). That estimate was lower than an estimate made by the Environmental Defense Fund (Denison 2005b) that called for an annual investment of $100 million per year by the federal government in nanomaterial-related EHS R&D. Denison (2005b, p. 1) acknowledged that "there is, of course, no single 'magic number' nor a precise means to determine the right dollar figure, given the wide-ranging set of research issues needing to be addressed and the significant associated uncertainty as to the anticipated results." Nevertheless, he proposed a rationale based on research gaps, nanotechnology R&D investment and market impact, and expert assessment and benchmarking, including the recommended and actual Environmental Protection Agency expenditures on airborne particulate-matter research.

Other organizations, such as the NanoBusiness Alliance (Murdock 2008), have argued that 10% of the federal nanotechnology-related R&D budget should be focused on EHS research. In testimony before the U.S. House of Representatives Committee on Science and Technology in 2008, Sean Murdock, executive director of the NanoBusiness Alliance, stated (Murdock 2008, p. 29):

> The NanoBusiness Alliance believes that environmental, health, and safety research should be fully funded and based on a clear, carefully-constructed research strategy. While we believe that 10 percent of the total funding for nanotechnology research and development is a reasonable estimate of the resources that will be required to execute the strategic plan, we also believe that actual resource levels should be driven by the strategic plan as they will vary significantly across agencies.

From 2006 to 2010, absolute and relative federal funding for nanotechnology-related EHS R&D increased substantially (see Table 5-1). In FY 2006, the federal government invested an estimated \$37.7 million in nanotechnology-related EHS R&D (2.8% of the nanotechnology R&D budget). In contrast, in FY 2010, nanotechnology-related EHS R&D accounted for 5.1% of the federal nanotechnology R&D budget, or \$91.6 million. In 2011, nanotechnology-related EHS funding showed a marked decrease; however, the President's FY 2012 budget request proposes \$123.5 million for nanotechnology-related EHS R&D—5.8% of the total nanotechnology R&D budget and the highest annual budget to date (NSET 2011). This figure does not include a substantial body of research on the biologic interactions and impacts of materials designed to enter the body for medical purposes – therapeutics, therapeutic delivery vehicles, and medical devices. While there is some disagreement over the direct applicability of research in this area to understanding the health and environmental risks associated with ENMs (NRC 2009), this is a research area that undoubtedly contributes to mechanistic understanding of how certain ENMs interact with biologic systems potentially causing harm. Therefore current overall federal investment in EHS-relevant R&D is likely substantially greater than \$123.5 million. However, it remains unclear how this investment translates into actionable information on potential EHS risks.

Although there has been concern that accounting of nanotechnology-related EHS R&D investment has been overinflated (GAO 2008; NRC 2009), the committee finds that current investments, as reported in the supplement to the President's federal budget request (NSET 2011), represent a reasonable indication of federal R&D funding specifically directed at EHS R&D. However, on the basis of the analysis of research needs presented in this chapter and in

TABLE 5-1 National Nanotechnology Initiative EHS Research Funding, FY 2006-2012

	Federal Nanotechnology-Related EHS R&D Investment	EHS Percentage of Total Federal Nanotechnology R&D Investment
FY 2006 (actual)	$37.7 million	2.8%
FY 2007 (actual)	$48.3 million	3.4%
FY 2008 (actual)	$67.9 million	4.4%
FY 2009 regular (actual)	$74.5 million	4.4%
FY 2009 ARRA[a] (actual)	$12.0 million	—
FY 2010 (actual)	$91.6 million	5.1%
FY 2011 (requested)	$116.9 million	6.6%
FY 2011 (continuing resolution)	$85.6 million	4.6%
FY 2012 (requested)	$123.5 million	5.8%

[a]The American Reinvestment and Recovery Act of 2009.
Source: Adapted from Sargent 2011; NSET 2010, 2011.

Chapters 3 and 4, the committee considers that there is still a gap between the research and associated activities currently funded and the level of activity that would foster greater and more responsive progress toward providing information and tools to support the safe development of nanotechnologies.

In general, the committee considers the predominant challenge to closing this gap is one of strategic realignment rather than additional funding. Based on the analysis conducted, the committee concludes that the research needs and research priorities addressed in this report provide an opportunity for strategically realigning the substantial federal resources being dedicated to ENM EHS R&D, based on the priorities outlined above. Such realignment will require federal-agency cooperation and resource leveraging.

In addition, modest resource increases in the five areas described below could have a substantial effect on building infrastructure that is critical for supporting an effective R&D program. The committee recognizes that such resource increases are not likely to be met by the budget requests of any one agency or institute but need to be garnered through a coordinated effort on the part of the nanomaterial community to leverage additional resources from public, private, and international initiatives to support critical cross-cutting research.

These critical cross-cutting activities are encompassed within the research priority categories described above and would be supported by greater coordinated investment in nanotechnology-related EHS informatics, investment in translating advanced measurement and characterization approaches to EHS-accessible methods, additional investment in developing and providing benchmark nanomaterials, investment in identifying and characterizing nanomaterial sources across the value chain and life cycle of products, and investment in developing and maintaining research networks that provide human infrastructure for collaborative research, information-sharing, and translation. Without budgetary increases in each of these areas, the committee anticipates that the federal government's ability to derive the maximum strategic value from investments in nanotechnology-related EHS research will remain insufficient.

Initial funding estimates made by the committee for these cross-cutting activities were based on anticipated costs of addressing specific challenges over the next five years, and where possible drew from similar examples from other fields (for instance, the development of informatics programs in areas outside nanotechnology). The results, however, were considered to be unrealistic within the current economic climate, representing an anticipated investment in excess of $200 million over the next five years. Instead, the committee considered from an expert perspective what levels of funding would enable substantial advances in research infrastructure, allowing significant value to be added to existing and emerging research programs, while working within realistic budgetary constraints. The resulting dollar amounts are based on expert judgment that is informed by the research priorities outlined above, the committee's grasp of the cost of research activities, and a consensus on the most appropriate balance between anticipated impact and realistic investment.

To ensure the development and implementation of a strategic nanotechnology-related EHS R&D program that is proportionate to overall nanotechnology R&D funding, that is commensurate with nanotechnology's economic and societal importance, that addresses critical knowledge gaps, and that maximizes the beneficial influence of federal R&D investments, the committee offers the following resource recommendations:

- On the assumption that core nanotechnology-related EHS R&D funding by federal agencies remains at about $120 million per year over the next 5 years, over time, funded research should be aligned with strategic priorities identified here and in the National Nanotechnology Initiative nanotechnology-related EHS strategy. Any reduction in this total would be a setback to EHS research and slow progress in addressing the committee's priorities.
- Additional multi-agency funding should be made available for five cross-cutting activities that are critical for providing needed infrastructure and materials to support a strategic and effective R&D program and for ensuring that research findings can be readily translated into practical action by stakeholders. The guidance on funding levels is general, and not indicative of priority order. The specified amounts are the minimums that should be available, and for each of the areas funding is critically needed in the short term:

- Informatics: \$5 million per year in new funding for the next 5 years should be used to support the development of robust informatics systems and tools for managing and using information on the EHS effects of ENMs. The committee concluded that developing robust and responsive informatics systems for ENM EHS information was critical to guiding future strategic research, and translating research into actionable intelligence. This includes maximizing the value of research that is EHS-relevant but not necessarily EHS-specific, such as studies conducted during the development of new therapeutics. Based on experiences from other areas of research, investment in informatics on the order of \$15 million is needed to make substantial progress in a complex and data rich field. However, within the constraints of nanotechnology R&D, the committee concluded that the modest investment proposed would at least allow initial informatics systems to be developed and facilitate planning for the long-term.
- Instrumentation: \$10 million per year in new funding for the next 5 years should be invested in translating existing measurement and characterization techniques into platforms that are accessible and relevant to EHS research and in developing new EHS-specific measurement and characterization techniques for assessing ENMs under a variety of conditions. The committee recognized that the proposed budget is insufficient for substantial research into developing new nanoscale characterization techniques—especially considering the cost of high-end instruments such as analytic electron microscopes—in excess of \$2 million per instrument. However, the proposed budget was considered adequate to support the translation of techniques developed or deployed in other fields for the EHS characterization of ENMs.
- Materials: Investment is needed in developing benchmark ENMs over the next 5 years, a long-standing need that has attracted little funding to date. The scope of funding needed depends in part on the development of public-private partnerships. However, to assure that funding is available to address this critical gap, the committee recommends that \$3-5 million per year be invested initially in developing and distributing benchmark ENMs. While more funds could be expended on developing a library of materials, this amount will assure that the most critically needed materials are developed. These materials will enable systematic investigation of their behavior and mechanisms of action in environmental and biologic systems. The availability of such materials will allow benchmarking of studies among research groups and research activities. The committee further recommends that activities around materials development be supported by public-private partnerships. Such

partnerships would also help to assure that relevant materials are being assessed.

- Sources: $2 million per year in new funding for the next 5 years should be invested in characterizing sources of ENM release and exposure throughout the value chain and life cycle of products. The committee considered that this was both an adequate and reasonable budget to support a comprehensive inventory of ENM sources.
- Networks: $2 million per year in new funding for the next 5 years should be invested in developing integrated researcher and stakeholder networks that facilitate the sharing of information and the translation of knowledge to effective use. The networks should allow participation of representatives of industry and international research programs and are a needed complement to the informatics infrastructure. They would also facilitate dialogue around the development of a dynamic library of materials. The committee concluded that research and stakeholder networks are critical to realizing the value of federally funded ENM EHS research and considered this to be an area where a relatively small amount of additional funding would have a high impact—both in the development of research strategies and in the translation and use of research findings. Given the current absence of such networks, the proposed budget was considered adequate.

REFERENCES

Auffan, M., W. Achouak, J. Rose, M.A. Roncato, C. Chaneac, D.T. Waite, A. Masion, J.C. Woicik, M.R. Wiesner, and J.Y. Bottero. 2008. Relation between the redox state of iron-based nanoparticles and their cytotoxicity toward *Escherichia coli.* Environ. Sci. Technol. 42(17):6730-6735.

Denison, R.A. 2005a. Getting Nanotechnology Right the First Time. Statement to the National Research Council Committee to Review the National Nanotechnology Initiative, 25 March 2005, Washington, DC [online]. Available: http://sei.nnin.org/doc/resource/balbus.pdf [accessed May 25, 2011]. [Pp. 172-174 in A Matter of Size: Triennial Review of the National Nanotechnology Initiative (2006). Washington, DC: National Academies Press].

Denison, R.A. 2005b. A Proposal to Increase Federal Funding of Nanotechnology Risk Research to at least $100 Million Annually. Environmental Defense Fund, Washington, DC [online]. Available: http://www.environmentaldefense.org/documents/4442_100milquestionl.pdf [accessed May 26, 2011].

EPA (U.S. Environmental Protection Agency). 2011. Inventory Update Reporting and Chemical Data Reporting [online]. Available: http://www.epa.gov/oppt/iur/ [accessed December 21, 2011].

GAO (U.S. General Accountability Office). 2008. Nanotechnology: Better Guidance is Needed to Ensure Accurate Reporting of Federal Research Focused on Environment, Health, and Safety Risks. GAO-08-402. Washington, DC: U.S. General Accountability Office. March 2008 [online]. Available: http://www.gao.gov/new.items/d08402.pdf [accessed May 9, 2011].

Gruber, T. 2011. What is an ontology? [online]. Available: http://www-ksl.stanford.edu/kst/what-is-an-ontology.html [accessed May 16, 2011].

Johnson, R.L., G. O. Johnson, J.T. Nurmi, and P.G. Tratnyek. 2009. Natural organic matter enhanced mobility of nano zerovalent iron. Environ. Sci. Technol. 43(14): 5455-5460.

Li, Z., K. Greden, P.J. Alvarez, K.B. Gregory, and G.V. Lowry. 2010. Adsorbed polymer and NOM limits adhesion and toxicity of nano scale zerovalent iron to E. coli. Environ. Sci. Technol. 44(9):3462-3467.

Maynard, A.D. 2006. Nanotechnology: A Research Strategy for Addressing Risk. PEN 3. Woodrow Wilson International Center for Scholars, Project on Emerging Nanotechnologies, Washington DC [online]. Available: http://www.nanotechproject.org/process/assets/files/2707/77_pen3_risk.pdf [accessed July 7, 2011].

Maynard, A.D. 2008. Annex A. Assessment of U.S. Government Nanotechnology Environmental Safety and Health Research for 2006. Testimony of Andrew D. Maynard, Chief Science Advisor, Woodrow Wilson International Center for Scholars the U.S. House of Representatives Committee on Science and Technology: The National Nanotechnology Initiative Amendments Act of 2008, April 16, 2008 [online]. Available: http://www.nanotechproject.org/process/assets/files/6690/maynard_written_april08.pdf [accessed May 9, 2011].

Murdock, S. 2008. Testimony of Sean Murdock, Executive Director, NanoBusiness Alliance before the U.S. House of Representatives Committee on Science and Technology: The National Nanotechnology Initiative Amendments Act of 2008, April 16, 2008 [online]. Available: http://gop.science.house.gov/Media/hearings/full08/april16/murdock.pdf [accessed May 9, 2011].

NEHI (Nanotechnology Environmental Health Implications Working Group). 2008. National Nanotechnology Initiative: Strategy for Nanotechnology-Related Environmental, Health, and Safety Research. Washington, DC: National Science and Technology Council. February 2008 [online]. Available: http://www.nano.gov/sites/default/files/pub_resource/nni_ehs_research_strategy.pdf?q=NNI_EHS_Research_Strategy.pdf [accessed July 8, 2011].

NRC (National Research Council). 2009. Review of the Federal Strategy for Nanotechnology-Related Environmental, Health, and Safety Research. Washington, DC: National Academies Press.

NSET (Nanoscale Science, Engineering, and Technology Subcommittee). 2010. The National Nanotechnology Initiative: Research and Development Leading to a Revolution in Technology and Industry. Supplement to the President's FY 2011 Budget. Washington DC: National Science and Technology Council [online]. Available: http://www.nano.gov/sites/default/files/pub_resource/nni_2011_budget_supplement.pdf [accessed July 8, 2011].

NSET (Nanoscale Science, Engineering, and Technology Subcommittee). 2011. The National Nanotechnology Initiative: Research and Development Leading to a Revolution in Technology and Industry. Supplement to the President's FY 2012 Budget. Washington DC: National Science and Technology Council [online]. Available: http://www.nano.gov/sites/default/files/pub_resource/nni_2012_budget_supplement.pdf [accessed May 9, 2011].

Phenrat, T., T.C. Long, G.V. Lowry, and B. Veronesi. 2009a. Partial oxidation ("aging") and surface modification decrease the toxicity of nanosized zerovalent iron. Environ. Sci. Technol. 43(1):195-200.

Phenrat, T., H.J. Kim, F. Fagerlund, T. Illangasekare, R.D. Tilton, and G.V. Lowry. 2009b. Particle size distribution, concentration, and magnetic attraction affect transport of polymer-modified Fe^0 nanoparticles in sand columns. Environ. Sci. Technol. 43(13):5079-5085.

Saleh, N.B., L.D. Pfefferle, and M. Elimelech. 2008. Aggregation kinetics of multiwalled carbon nanotubes in aquatic systems: Measurements and environmental implications. Environ. Sci. Technol. 42(21):7963-7969.

Sargent, J.F. 2011. Nanotechnology and Environmental, Health, and Safety: Issues for Consideration. Congressional Research Service Report for Congress. January 20, 2011 [online]. Available: http://www.fas.org/sgp/crs/misc/RL34614.pdf [accessed May 9, 2011].

6

Implementing the Research Strategy and Evaluating Progress

INTRODUCTION

This report has articulated the rationale for an environmental, health, and safety (EHS) risk-research strategy related to engineered nanomaterials (ENMs) (Chapter 1) and proposed a conceptual framework for addressing EHS risks (Figure 2-1). Chapter 3 summarized the current state of the science and high-priority data gaps on the potential EHS risks posed by ENMs. Chapter 4 described the fundamental tools and approaches needed to pursue an EHS risk research strategy. Chapter 5 presented the committee's proposed research agenda, short-term and long-term research priorities, and estimates of needed resources.

This chapter focuses on implementation of the research strategy and evaluation of its progress, elements that the committee considered integral to its charge. It envisions a strategy that is fully integrated and coordinated, that addresses short-term and long-term needs, and that evolves as information is gleaned and progress is made. It places the discussion of Chapters 3-5 into the context of this broader vision. The present chapter also addresses the following core components that reflect elements identified in the 2009 National Research Council report (NRC 2009) as necessary for implementing an effective research strategy:

Infrastructure for implementation and accountability: Two distinct types of new or expanded infrastructure are needed for implementing the strategy: expansion of institutional arrangements, including interagency coordination, stakeholder engagement, public-private partnerships, and management of potential conflicts of interest; and mechanisms for integrating informatics and information-sharing into the research structure.

Evaluation of research progress and revision of the strategy: A critical element of an effective research strategy is the inclusion of processes for evaluation of progress in relation to the goals of the strategy and for refinement and adaptation of the strategy as the information base evolves and drives the need for change.

Resources needed to conduct research and implement the strategy: An integral part of an effective strategy is a means for continuing assessment of resources for conducting the research and for implementing the strategy. That includes assessing the adequacy of current resources, how they are to be deployed, and how additional resources would best be acquired and used.

Key audiences for implementing the strategy: The strategy should effectively consider and integrate the needs of and appropriate roles for the full array of stakeholders involved in implementing the strategy or concerned with its outcomes. These stakeholders include the National Nanotechnology Initiative (NNI) and the federal agencies within its structure; the private sector, including nanomaterial developers and users; consumers; and the broader scientific community, including academic researchers and non-governmental organizations.

Many of the questions posed in formulating a strategy for research on nanomaterials are equally applicable to strategies that might be developed for other emerging technologies. Lack of knowledge and high degrees of uncertainty, with a rapidly changing landscape of perceived risks and benefits, are inherent in the rollout of any new technology. As with nanomaterial research, there are likely to be challenges in developing common protocols for a community of researchers that turns its attention to the evaluation of the environmental and health implications of any new technology and seeks to compare the results of research among laboratories. For those reasons, the proposed research strategy can be considered a model approach for developing a strategy that examines the risks produced by other emerging technologies.

INFRASTRUCTURE FOR IMPLEMENTATION AND ACCOUNTABILITY

Mechanisms for effective implementation of an EHS research strategy are just as essential to its success as is the substance of the identified research, a key finding in NRC (2009). Questions that must be grappled with include the following: What are the most effective approaches to achieving the stated goals? How will research efforts be coordinated to ensure a coherent approach to achieving the stated goals? What mechanisms and approaches are in place or could be created to enable interdisciplinary research that crosses established funding and agency boundaries and to foster effective coordination and partnerships?

As noted in Chapter 1, the committee acknowledges the contribution that the NNI has made toward implementation (NEHI 2010[1], Chapter 8). Most nota-

[1]A final version of the strategy was published in October 2011 (NEHI 2011a). Because the committee's report had already gone to peer review, NEHI (2011a) was not reviewed by this committee.

bly, through the naming of the National Nanotechnology Coordination Office (NNCO) Coordinator for EHS, the Office of Science and Technology Policy (OSTP) has vested leadership and a measure of accountability for EHS research. This NNCO EHS coordinator is charged with leading efforts to integrate nanotechnology EHS research, to identify and leverage domestic and international collaborations, and to serve as the NNI point of contact for stakeholders on nanotechnology EHS issues (NEHI 2010). The Nanotechnology Environmental Health Implications (NEHI) working group (NEHI 2010) also calls on NNI agencies to explore and exploit new media and networking opportunities to improve interagency communication and stakeholder interaction, to develop new mechanisms for NNI agencies to partner with industry, to facilitate "development of joint programs among NNI agencies to fund EHS research of mutual interest and avoid unproductive redundancy" (p. 78), to expand work in data storage and management, and to "develop and maintain data communication infrastructures and organization" (p.79).

The committee lays out below implementation needs of two major kinds: institutional arrangements and informatics and information-sharing. The discussion includes models and means to address the key needs and provide accountability to the broader community of stakeholders.

Institutional Arrangements

As noted earlier, to ensure successful implementation of a research strategy, accountability must be integral to the strategy's development and execution. There are numerous dimensions of accountability:

- Ensuring and assigning ownership of the overall strategy.
- Establishing appropriate means of governance among parties implementing the strategy.
- Establishing and applying mechanisms for accomplishing exploratory, translational, and targeted research in the context of the strategy, including an appropriate balance between government and private-sector funding and facilitating needed or desired interdisciplinary research.
- Assigning responsibility for executing elements of the strategy.
- Ensuring that stakeholders are involved in and have substantive input into formulating and reviewing the strategy.
- Monitoring progress in comparison with elaborated research goals and timelines to ensure that the strategy is conducted effectively and efficiently and to ensure that responsible parties are held accountable for the extent of research progress.
- Coordinating periodic review and revisions to the strategy.
- Ensuring that sufficient resources are devoted to conducting the needed research and to implementing the overall strategy and allocating and managing the resources.

• Managing potential or perceived conflicts of interest resulting from the dual missions of the federal government in nanotechnology—investment in the development and commercialization of nanotechnology and in ensuring its safety.

• Ensuring wide dissemination of research results to decision-makers and other stakeholders.

This section discusses the needs for and approaches to providing accountability in implementing the strategy in four categories: enhancing interagency coordination, providing for stakeholder engagement in the research strategy and its revision, conducting and communicating the results of research funded through public-private partnerships, and managing potential conflicts of interest. The committee's conclusions in the four areas also are presented.

Enhancing Interagency Coordination

A shortcoming identified in the 2009 National Research Council report was the inadequacy of mechanisms in the NNI structure to ensure accountability for implementation of the research strategy. This problem reflects the largely coordinating and information-sharing roles of the NNI and of the primary body in the NNI that addresses environmental health and safety issues, the Nanoscale Science, Engineering and Technology (NSET) Subcommittee's NEHI. NEHI's mission includes (NEHI 2011b)

• "Providing for *information exchange among Federal agencies* that support nanotechnology research . . .;

• "*Facilitating* the identification, prioritization, and implementation of research and other activities required for the responsible research necessary to develop, use, and oversee nanotechnology;

• "*Promoting communication of information* related to research on environmental, health, and safety implications of nanotechnology;

• "Adaptively managing (i.e., *coordinating*, reviewing, and revising) the interagency EHS research strategy . . .;

• "*Assisting in developing information and strategies* as a basis for drafting guidance in the safe handling and use of nanomaterials and products;

• "With input from NSET and other interagency groups, *supporting the development of tools and methods* to identify, prioritize, and manage strategies for specific research to enable risk analysis and regulatory decision-making for nanomaterials and products incorporating nanomaterials [emphasis added]."

Those "bottom-up" functions—interagency coordination, information exchange, facilitation, and communication—clearly are important and necessary. However, they are insufficient to ensure the degree of accountability needed to develop and execute a national EHS research strategy (NRC 2009, pp. 47-49).

NEHI's function needs to be supplemented with a "top-down" authority to direct and oversee the EHS research budgets and associated activities within and among NNI agencies and between government and the private sector (Denison 2007a, p. 4).

As discussed previously, OSTP's naming of an NNCO EHS coordinator is a step in this direction. However, as the title of the position suggests, the role of the coordinator stops short of the needed "top-down" authority inasmuch as the role is to "ensure effective communication and coordination of the NNI . . . and of agency EHS R&D efforts and integration of these efforts with the NNI Strategy for Nanotechnology-Related EHS Research (Nanowerk News 2010)." Hence, it appears that the new position principally formalizes and raises the role and function already served by NEHI rather than creating an entity with additional budget and oversight authority.

Various models have been proposed and others might be considered to provide greater authority. One model envisioned under the proposed "NNI Amendments Act of 2010" (Section 103 of Subtitle A of H.R. 5116, the America COMPETES Reauthorization Act of 2010)[2] was for an associate director of OSTP to be appointed and made responsible for "oversight of the coordination, planning, and budget prioritization" of EHS research-related activities. The act further delineated those responsibilities:

> "(1) ensuring that a research plan for the EHS research activities required under subsection (b) is developed, updated, and implemented . . .,
> "(2) encouraging and monitoring the efforts of the agencies participating in the Program to allocate the level of resources and management attention necessary to ensure that the . . . environmental . . . concerns related to nanotechnology, including human health concerns, are addressed under the Program, including the implementation of the research plan described in subsection (b); and
> "(3) encouraging the agencies required to develop the research plan under subsection (b) to identify, assess, and implement suitable mechanisms for the establishment of public-private partnerships for support of EHS research."

Under the proposed legislation, the OSTP associate director would chair a panel that comprised senior officials of the federal agencies funding relevant research "to develop, periodically update, and coordinate the implementation of a research plan." In carrying out that activity, the associate director would be

---

[2]This passed the U.S. House of Representatives on May 29, 2010. S. 3605, the Senate bill of the same name, was reported out of the Committee on Commerce, Science, and Transportation, but it did not include the provisions related to nanotechnology that were in the House bill. It was the Senate version of the America COMPETES Reauthorization Act that ultimately was enacted in the 111th Congress and signed by President Obama on January 4, 2011.

required to solicit and take into account the recommendations of a newly established EHS subpanel of the external advisory panel established pursuant to the original legislation that established the NNI (the 21st Century Nanotechnology Research and Development Act of 2003, Section 4).

The research plan called for under the proposed legislation would have been required to:

> "(A) specify near-term research objectives and long-term research objectives;
> "(B) specify milestones associated with each near-term objective and the estimated time and resources required to reach each milestone;
> "(C) with respect to subparagraphs (A) and (B), describe the role of each agency carrying out or sponsoring research in order to meet the objectives specified under subparagraph (A) and to achieve the milestones specified under subparagraph (B);
> "(D) specify the funding allocated to each major objective of the plan and the source of funding by agency for the current fiscal year; and
> "(E) estimate the funding required for each major objective of the plan and the source of funding by agency for the following 3 fiscal years."

The plan was required to be made public and updated annually. A public database was to be established and maintained with all EHS research projects funded under the plan by NNI agencies, "including a description of each project, its source of funding by agency, and its funding history, . . . grouped by major objective as defined by the research plan." (See Section 102 of Subtitle A of H.R. 5116, the America COMPETES Reauthorization Act of 2010.)

Most of the previously identified elements to ensure accountability would be addressed through the provisions of the 2010 proposed legislation. The proposed legislation was passed by the House of Representatives but not by the Senate in the 111th Congress; at the time of this writing, it has not been reintroduced in the 112th Congress.

Another approach that might be housed in the existing NNI and NEHI structure would be to designate a person (Maynard 2007), a small group of senior health and environmental officials (Denison 2007b), or an agency with lead responsibility and to provide this entity with the budgetary and management authority needed to direct the EHS research. The officials might be drawn principally from NNI agencies whose missions are to protect human health and the environment and have related research capabilities. Whether situated in or outside the current NNI structure, such an entity would need to have decision-making authority that is independent of the parts of NNI charged with advancing nanotechnology development (Denison 2007a, p. 4). (See section below, "Managing Potential Conflicts of Interest.")

An example that may serve as a model for interagency coordination and sharing of research roles is the agreement that was reached between the Environmental Protection Agency (EPA) and the Department of Energy (DOE) concerning radon research (EPA 1987, 1989). In the middle 1980s, extremely high concentrations of radon were found in homes in Pennsylvania and led to intense public interest and Congressional attention to issues related to indoor radon. In a reauthorization of the Superfund Amendments and Reauthorization Act, Congress designated EPA as the lead agency for addressing radon and other indoor pollutants. However, DOE had a long history of research in radiation biology, anthropogenic sources of radiation exposure, and energy efficiency and building science. Each agency had its supporters, and the tension between them created considerable turmoil.

To address the problems, a memorandum of understanding was signed in September 1987 that defined primary responsibilities for each agency (EPA 1987, 1989). EPA's role was applied research, particularly that related to monitoring and mitigation techniques and to operational indoor-radon programs. DOE was responsible for basic research in methods for reducing overall exposure to radiation, for investigation of health and environmental effects of radon exposure, and for issues related to the further development of energy efficiency in buildings. Within a few years, EPA clearly led with action programs and public outreach regarding radon exposures, and DOE had a substantial basic-research and applied-research program that fostered understanding of exposure issues, radiation biology, and basic science.

Another model of coordination among federal agencies originates in OSTP: the Committee on Environment, Natural Resources, and Sustainability (CENRS) established by the National Science and Technology Council (OSTP 2011). CENRS is cochaired by OSTP, the National Oceanic and Atmospheric Administration, and EPA. It provides a formal mechanism for interagency coordination relevant to environment, natural resources, and sustainability policy and R&D issues nationwide and globally. The model has been effectively used in coordinating research on particulate matter (PM) funded by multiple federal agencies. In the early 1990's, PM research was largely uncoordinated and fragmented. Following the promulgation of a new PM standard by EPA in 1997, there was a call for a research agenda. A report by the NRC Committee on Research Priorities for Airborne Particulate Matter published in 1998 laid the groundwork for an integrated research and management program. This committee was charged with monitoring progress and did so over a six-year period. Within a few years, coordination among funding agencies, investigators, and regulators benefited from an interagency working group chaired by EPA but with representation of all the key participants (CENR 2011). The oversight effort identified research gaps and collectively supported the research to address them. The oversight program also fostered collaborative efforts among investigators when this approach was perceived as the optimal means of solving problems. The coordination of the PM research agenda clearly has enhanced research productivity and utility in public-policy decisions. CENRS also is using the model to coordinate climate issues in the federal government, including upgrades of meteorology and air-quality models and how the improved models should be optimized for global and regional issues.

NEHI (2010) acknowledges the need for stronger agency coordination and describes some new initiatives, including more active involvement by the Consumer Product Safety Commission and the Food and Drug Administration, the naming of the new NNCO EHS coordinator, the development of an interagency Web site, and a new charge to NEHI to clarify research priorities and identify cross-agency collaboration opportunities.

Conclusion:

While recognizing the important coordinating role of NNI, the committee concludes that to implement its strategy effectively an entity with sufficient management and budgetary authority is needed to direct development and implementation of a federal EHS research strategy throughout NNI agencies and to ensure its integration with EHS research undertaken in the private sector, the academic community, and international organizations. Progress in implementation of the strategy will be severely limited in the absence of such an entity. It would be helpful for NNCO to identify funding needs and mechanisms for interagency collaboration on high-priority research annually.

Providing for Stakeholder Engagement in the Research Strategy

NRC (2009) concluded that the federal nanotechnology-related EHS research strategy did not adequately seek and incorporate the inputs and perspectives of various stakeholders. Input into the strategy was limited to that from the NNI agencies and was constrained by their experience, expertise, and resources. The review concluded that such an insular approach results in an insular research strategy (NRC 2009, p. 49):

> The reason is that federal agencies have a vested interest in justifying the applicability of current efforts rather than critically assessing what is not being done and how deficiencies might be addressed. For example, when agencies are developing their own research strategies, they tend to ask, What research can we do within our existing capabilities?, rather than the more appropriate, What research should we be doing? Other relevant questions need to be addressed, such as, Are resources adequate? Are adequate mechanisms and organizational structures in place to achieve the desired goals? As a result, the federal strategy becomes a justification for current activities based on a retrospective examination that demonstrates success rather than the development of a prospective strategy that questions current practices with an eye to future research needs.

Only by actively soliciting from the outset and integrating the needs of those who have a stake in the outcomes of the strategy can it be responsive and ensure that the right questions are being asked and answered.

Progress has been made in engaging stakeholders in the development of NEHI (2010). NNI held four public workshops focused on various aspects of nanotechnology-related EHS that solicited reaction to the 2008 strategy (NEHI 2008, 2010): "Human and Environmental Exposure Assessment", "Nanomaterials and the Environment and Instrumentation, Metrology, and Analytical Methods", "Nanomaterials and Human Health and Instrumentation, Metrology, and Analytical Methods", and "Risk Management and Ethical, Legal, and Societal Implications of Nanotechnology." Participants were asked to review and update the state of the science and to identify critical research gaps and barriers to conducting needed research.

NNI also published a request for information in the *Federal Register* and established an online portal to receive input on its 2011 draft strategy during a designated comment period (NEHI 2010).

The NNI has established several industry partnerships called Consultative Boards for Advancing Nanotechnology (CBANs) that are limited to industry and, as the name suggests, are aimed primarily at developing nanotechnology. Nonetheless, they do provide a potential means of obtaining input from industry stakeholders.

NRC (2009) cited one model for stakeholder input drawn from the practice of an existing NNI agency, the National Institute for Occupational Safety and Health (NIOSH) (NORA 2008):

> The National Occupational Research Agenda (NORA) is a partnership program to stimulate innovative research and improved workplace practices. Unveiled in 1996, NORA has become a research framework for NIOSH and the nation. Diverse parties collaborate to identify the most critical issues in workplace safety and health. Partners then work together to develop goals and objectives for addressing these needs.

NORA has several appealing features that go beyond NNI's efforts to date and that could be considered as a model for stakeholder engagement. NORA provides means by which the needs of stakeholders—"from universities, large and small businesses, professional societies, government agencies, and worker organizations"—inform the research questions and by which stakeholders are directly involved in the development of the research strategy designed to meet those needs. Various means of involvement are provided, recognizing the differential capacity of individuals and stakeholder groups.

In contrast with the process used by NNI, the NORA process of stakeholder engagement is continuing, rather than one-time or occasional, with input solicited and considered at all stages of strategy development and execution. Finally, national and sector-specific research agendas are produced and maintained in addition to an overall national strategy.

NORA identifies the diversity of stakeholders that it engages as key to its success. Going well beyond submitting comments on draft documents or attending meetings, stakeholders actively participate in standing "councils." For example, the Manufacturing Sector Council has representatives of state agencies, hospitals, insurance companies, universities, consulting firms, federal agencies, manufacturers, and labor unions. In addition to producing its sector research agenda, the council prepared and released a dozen fact sheets summarizing the state of information for different occupational health and safety problems in the manufacturing sector. Leaders of each sector council participate in a cross-sector council, the goal being "to enhance the effectiveness of the NORA Sector Councils through coordination of their activities, sharing information, and seeking efficiencies in dealing with issues that are common in two or more sectors" (NORA 2011).

The standing nature of the NORA stakeholder venues, the variety of interim and final products, the sector-specific strategies to supplement the national strategy, and the diversity of stakeholders engaged are all features that could be incorporated into a stakeholder-engagement effort focused on nanotechnology research needs and strategy development. NORA holds biennial symposia to discuss research findings, their implications, and future needs.

Although NORA serves the needs of NIOSH in soliciting stakeholder engagement, NNI's purview is far broader and involves multiple agencies and departments with differing agendas, cultures, processes, and practices. The NORA model would need considerable adaptation to ensure meaningful stakeholder engagement in this more complex setting. Nonetheless, some of NORA's features—opportunities to provide input, the provision for substantive stakeholder leadership roles, and the success in involving a broad array of stakeholders—would be desirable features of stakeholder engagement needed to develop and implement the current research strategy.

Conclusions:

To implement its strategy effectively, the committee concludes that several elements are needed to enhance stakeholder engagement:

- Establishing a standing means to receive input and actively engage stakeholders at all stages of strategy development, implementation, and revision.
- Establishing stakeholder groups representing or with interest in EHS implications relevant to specific sectors of application of nanotechnology and nanomaterials.
- Ensuring that diverse stakeholders are provided with a means of playing leadership roles in strategy development, implementation, and revision.

The committee notes that these conclusions go beyond stakeholder engagement opportunities provided under the NNI, and provide a framework for engagement of a broader cross-section of stakeholders. For example, standing councils and stakeholder groups need not and should not be limited to "inside-the-beltway" participants. The process for selecting their membership should be designed to be as inclusive as possible and invite nominations through broadly disseminated channels.

Conducting and Communicating the Results of Research Funded Through Public-Private Partnerships

NEHI (2010) briefly acknowledges the need to develop and leverage public-private partnerships. It notes the continuing use of the Small Business Innovation Research program and the Small Business Technology Transfer program. It notes that "new programs could support industry-public partners-agency collaborations on EHS research," but it provides or proposes none. NEHI (2010) has an equally short and nonspecific section on the related topic of knowledge dissemination.

As noted earlier, the NNI has established several industry partnerships called Consultative Boards for Advancing Nanotechnology (CBANs), though these are, as the name suggests, aimed primarily at developing nanotechnology.

Another example of a partnership is the Industry Consortium for Environmental Measurement of Nanomaterials (ICEMN) that involves stakeholders from industry and academia who are working to provide the California Department of Toxic Substances Control (DTSC) or other regulatory bodies with information that could be used to measure nanomaterials in air, surface water, and soil and to assess if these methodologies can be adapted to quantify or to characterize environmental concentrations of these nanomaterials. The consortium was established in response to a "data call-in" by DTSC for information on analytic test methods, and on fate and transport in the environment from manufacturers or importers of certain nanomaterials (DTSC/ICEMN 2011; NIA 2011).

Various models have been recommended by stakeholders to facilitate effective, reliable public-private partnerships to conduct nanomaterial EHS research. One such model is the Health Effects Institute (HEI) (HEI 2011). HEI is a nonpartisan, nonprofit research institute, jointly funded by the automotive industry and EPA that focuses on the health effects of air pollution. Strengths of the HEI model include its ability to solicit and fund targeted, policy-relevant research, its quality control and independent governance and peer-review procedures, and its operational transparency and commitment to release all research results (Denison 2007a; Maynard 2007). Others have cited limitations of the model for nanotechnology-related EHS research, including a much broader scope of research than that under the HEI, challenges associated with conducting research on proprietary materials, and research that occurs in many different agencies and industry sectors (Teague 2007).

Other models that have been considered germane to the conduct and communication of nanotechnology-related EHS research through public-private partnerships include the Foundation for the National Institutes of Health (FNIH 2011) and the National Cancer Institute (NCI) Nanotechnology Characterization Laboratory (NCI 2011a).

Conclusion:

To implement its strategy effectively, the committee concludes that EHS research requires public-private partnerships that provide for quality control, independent governance, peer-review of research results, operational transparency, and a commitment to release all research results and underlying data.

Managing Potential Conflicts of Interest

NNI was established to fill dual functions: to develop and promote nanotechnology and its applications and to identify and mitigate risks arising from such applications (21st Century Nanotechnology Research and Development Act [2003]). That duality is reflected in the diverse missions of the agencies and departments that make up NNI and in some of the offices within agencies engaged in NNI activities.

The housing of dual purposes in the same entity (NNI) and their coordination through the same office (NNCO) have raised concerns among some stakeholders regarding the potential for a conflict of interest. NRC (2009) noted that the conflict is a "false dichotomy," but it is clear that there are tensions between the goals of the two functions. The debate over the adequacy of the portion of the total NNI budget devoted to addressing EHS concerns is one manifestation. An equally contentious disagreement is that over NNI's classification of research projects with respect to their "EHS relevance" and its tendency to "overcount" the dollars spent on EHS research (NRC 2009). NRC (2009, p. 11) noted that "the committee is concerned that the actual amount of federal funding specifically addressing the EHS risks posed by nanotechnology is far less than portrayed in the NNI document and may be inadequate." The present committee is pleased that the accounting for and reporting of direct EHS-research spending are much improved in the latest draft NNI strategy (NEHI 2010).

In response to those concerns, NRC (2009) and stakeholders have called for more distinct lines of authority to be established to perform the two functions. The NRC (2009, p. 11) committee stated that "a clear separation of accountability for development of applications and assessment of potential implications of nanotechnology would help to ensure that the public-health mission has appropriate priority." One stakeholder, the Environmental Defense Fund, provided specific recommendations for achieving this separation in accountability (Denison 2007b, p. 7):

> Ensuring that both goals receive equal consideration would require, at a minimum, that the responsibility to address the two distinct goals be assigned to different offices and senior staff members, who are given parallel and comparable degrees of authority, and who report directly to the highest levels within their individual agencies and within NNI. We believe

that a clear division of labor and interests is critical if public confidence in the ability of the federal government to facilitate the responsible development of nanotechnology is to be restored.

Such a structure would have the additional advantage of ensuring that risk implications of nanotechnology are addressed by research that is intended and directly targeted to answer specific risk-relevant questions and that such research would be directed by—and its relevance and adequacy assessed by—scientists trained in the health or environmental sciences who work in agencies charged with these related missions (Denison 2007a, p. 4). It also would facilitate more transparent accounting of the funding of direct risk research, as distinct from more basic or applications-oriented research, some of which may also yield findings relevant to understanding risk (Denison 2007a). A separation in authority and accountability need not and should not be accomplished in a manner that would "silo-ize" risk research or impede cross-fertilization and synergies between the two lines of research or the free flow of research ideas and results.

A historical precedent and model for addressing perceived and actual conflicts between the federal government's interest in developing and its interest in managing the risks posed by new technologies was the establishment of the Nuclear Regulatory Commission (U.S. NRC). The Atomic Energy Commission (AEC), established by the Atomic Energy Act of 1946, was explicitly assigned the functions of both encouraging the use of nuclear power and regulating its safety. Concerns about that dual charge grew among proponents and critics of nuclear power and came to a head in the middle 1970s, when Congress abolished AEC. Congress then assigned the oversight functions of AEC to a new entity, U.S. NRC, and shifted federal nuclear energy research and development to DOE (Denison 2007a, p.4; U.S. NRC 2010a,b).

U.S. NRC's mission and work specifically include risk research: "As part of its regulatory program, the NRC conducts an extensive research program to provide independent information and expertise to support its safety decision making" (U.S. NRC 2010a). That research is conducted through the U.S. NRC's Office of Regulatory Research, which "provides leadership and plans, recommends, manages and implements programs of nuclear regulatory research" (U.S. NRC 2010c). The office also engages in considerable cooperative research with "DOE and other federal agencies, the nuclear power industry, U.S. universities, and international partners" (U.S. NRC 2010c). However, it operates and is managed independently, and U.S. NRC has extensive guidelines and procedures intended to ensure that it avoids conflicts of interest that could arise from its use of DOE laboratories for technical assistance and research (Callan 1998; Travers 1999) or from its hiring contractors who have also worked on or are competing for DOE contracts (Dingbaum 2002; Denison 2007a).

Far from operating in a "silo" and being unable to take advantage of the cross-fertilization arising from research conducted on applications, U.S. NRC has established an approach intended to allow safety research to be conducted in a manner that transparently manages potential conflicts of interest while maintaining its independent decision-making (Denison 2007a, p. 5). Adoption of a similar set of accountability mechanisms for nanotechnology-related EHS research would help to ensure that nanotechnology's risk implications get the attention that they need even as federal investment in nanotechnology development proceeds (Denison 2007a, p. 1).

Conclusions:

To implement its strategy (or other strategies) effectively, the committee concludes that a clear separation in management and budgetary authority and accountability is needed between the functions of developing and promoting applications of nanotechnology and of understanding and assessing its potential health and environmental implications. Such a separation is needed to ensure that progress in implementing an effective EHS-research strategy is not hampered. The separation in management of applications-targeted and implications-targeted research needs to be achieved through means that do not impede the free flow of ideas and results between the two lines of research.

To that end it would be helpful for:

- Responsibility in addressing the two distinct goals to be assigned within NNI agencies to different offices and senior staff members, who would be given parallel and comparable degrees of authority and would report directly to the highest levels in their individual agencies and in NNI.
- Research that is directly targeted at understanding risk questions to be tracked and reported separately from other more basic or applications-oriented research even though the latter may well yield findings relevant to understanding risk.
- The targeting and assessment of the relevance and adequacy of risk-relevant research to be assessed by scientists trained in the health or environmental sciences who work in agencies charged with health and environmental protection (Denison 2007a, p. 4).

Informatics and Information-Sharing

Just as institutional arrangements are critical to the implementation of the nanotechnology-related EHS research strategy, so is the development and use of informatics for information collection, analysis, and sharing. Chapter 4 described informatics in the context of method and model development and validation and presented research needs in those activities. Chapter 5 addressed informatics as one component of a larger research and knowledge infrastructure needed to advance the committee's research priorities. This section addresses broader organizational means of enhancing the collaboration necessary for implementation of an informatics infrastructure. Appendix B presents implementation scenarios for the development of methods, predictive models, a federated data-sharing network, and a semantic informatics infrastructure to illustrate how a systems approach could accelerate nanoscience and nanotechnology research and translation.

Informatics depends on the acquisition, processing, and sharing of large amounts of data and models (NNN 2011). Optimal use of informatics requires collaboration among academics, industry, regulatory bodies, metrology institutions, and laboratories. It also requires working relationships among various organizations in the research community—standard-development organizations, contract research organizations, material providers, and organizations performing interlaboratory studies. The benefits of collaboration are numerous and include the breaking down of data silos, the ability to conduct semantic searches and to share data and models, the use of Web-based tools for rapid dissemination and communication between disciplines, and ultimately acceleration of research (NEHI 2010). There have been not only scientific and technical barriers to broader use of informatics but organizational and cultural challenges. That is evidenced in part by the fact that despite the large amount of nanotechnology-related data that have been produced over the last 10 years in academic and industry laboratories, there remains a dearth of reliable, discoverable data that are standardized, verified, and capable of being shared effectively (NNN 2011, p. 13).

Issues that stymie collaboration are varied and include cultural barriers to data-sharing, intellectual-property concerns regarding data-sharing and data access, differences in expectations, unmet needs for proper annotation and attribution, lack of incentives, and differences in technologic infrastructure. The discussion below provides examples of means by which collaboration may be fostered to support an informatics infrastructure to aid nanotechnology-related EHS research.

One means of facilitating coordination of informatics projects is the Semantic Web, which is a set of practices and standards designed to enable individuals to structure their data so that they are compatible with Web-based exchange. The Semantic Web provides a common framework for data-sharing and data reuse among applications, enterprises, and communities (NEHI 2010; NNN 2011, p. 23). Ontologies have evolved for specific sets of users that have common naming conventions ("namespaces") and allow computers to search similar concepts to identify appropriate data regardless of how they are organized in a given namespace. For example, data from the Gene Ontology (used for mining genetic data) (Gene Ontology 2011) can be combined with data from the Nanoparticle Ontology (which integrates data in nanomedicine) (Thomas et al. 2011).

The power of the Semantic Web for nanotechnology is that it allows separate database systems to share their data and basic applications so that they are interoperable and can be easily joined ("federated") through a common framework for data-sharing. Such a system avoids the proliferation of data silos while allowing data to be annotated, curated, and maintained by experts. That flexibility even provides for international data-sharing in spite of language differences.

One model of collaboration that relies on the Semantic Web is NCI's cancer Biomedical Informatics Grid (caBIG). caBIG aims to create a collaborative computational and research network that connects scientists and institutions to facilitate collaboration, data integration, and data-sharing in cancer research (Fenstermacher et al. 2005). The NCI caBIG Nanotechnology Working Group (caBIG Nano WG)—which comprises participants in academe, government agencies, industry, and other organizations—was established in 2009 for researchers interested in applying informatics and computational approaches to nanotechnology, with an emphasis on nanomedicine. caBIG Nano WG is now integrating data, federating nanotechnology databases via pilot projects for enabling the semantic search and retrieval of nanomedicine and nanotoxicology datasets, and aiding in the dissemination of standard protocols (NNN 2011, p. 17). For example, a pilot portal (the cancer Nanotechnology Laboratory [NCI 2011b]) is federating with other databases—such as the Oregon Nanoscience and Microtechnologies Institute Nanomaterial Biological Interactions knowledge base, the National Nanomanufacturing Network Web portal, the NIOSH Nanoparticle Information Library, NanoHUB, and the Collaboratory for Structural Nanobiology—in a series of demonstrations (Nanoinformatics 2010). The demonstrations serve as test beds to elicit requirements from stakeholders for future collaborations in the development, validation, and dissemination of analytic methods; information on the error, uncertainty and robustness of the methods; laboratory expertise; the minimum characterization required for particular ENMs; the development, validation, and sharing of structural, predictive, and risk models; and access to computational facilities.

Grass-roots initiatives that are intended to coordinate networks of researchers could help to foster collaboration in the collection, curation, dissemination, and analysis of nanotechnology-related EHS data and to engage stakeholders. One example outside the nanotechnology realm is the iPlant Collaborative (iPlant Collaborative 2011), a $50 million-per-year program funded by the National Science Foundation (NSF) and begun in 2008 at the University of Arizona to foster communication and share content in the plant-science community.

The intent of sharing data and making data accessible stems from the principle that doing so will move science forward (NAS/NAE/IOM 2009). The practice is becoming more standard. For example, NSF now requires all grant proposals to include a two-page data-management plan for how data will be disseminated and shared publicly (NNN 2011, p. 19). (See Box 6-1 for additional discussion.) Such efforts could provide opportunities for developing formats and databases for nanomaterial data and further encourage collaboration.

BOX 6-1 National Science Foundation Data-Management Plan

One strategy for developing a data source and standard submission practice is NSF's recently instituted requirement for a data management plan: proposals submitted or due on or after January 18, 2011, must include a supplementary "Data Management Plan" that describes how data will be disseminated and shared publicly. There is no repository for this information or standard format for submission; NSF merely requires that data be made available in some form, be archived, and include "analyzed data and the metadata that define how these data were generated. These are data that are or that should be published in theses, dissertations, refereed journal articles, supplemental data attachments for manuscripts, books and book chapters, and other print or electronic publication formats."

The "Data Management Plan" requirement, combined with the fact that each directorate has its own requirement for deposition of information and with the current lack of a defined structure, provides an excellent opportunity to develop formats and databases for nanomaterial data that could be standardized for programs that involve nanomaterial development or EHS research. NSF could be directed to lead the effort to develop such databases, given its data-management requirement.

Conclusion:

The committee considers that NNI has taken an important step in identifying and committing to implementation of an informatics system in its 2010 draft research strategy (NEHI 2010). However, to address the high-priority research needs outlined in Chapter 5, mechanisms are needed to facilitate nanotechnology-related EHS research in the larger community of researchers and decision-makers beyond NNI. With input from the community of researchers and other stakeholders, the mechanisms may include incentives or even requirements for data-sharing, funding to develop repositories, and resources for community-building to help government and nongovernment researchers to make needed connections among disciplines with colleagues worldwide.

EVALUATING AND ASSESSING PROGRESS FOR REVISING THE STRATEGY

To be relevant, timely, and effective, the development and implementation of a risk-research strategy is an iterative process. It begins with research planning and requires focused, creative, and flexible management of the implementation process. It necessarily includes a process for monitoring and evaluating the agenda's progress in generating the scientific research and policy-relevant information needed to reduce uncertainty and to address high-priority scientific-knowledge and decision-making gaps. That in turn informs a review process that may result in updates, revisions, and adjustments of the research agenda.

Review and Adaptation

Research activities require periodic reassessments to maintain focus, relevance, and accountability. Current activities need the same level of review and scrutiny as newly initiated activities to ensure maximum return on research investments (NRC 1998, p. 118). That is particularly true of the dynamic and rapidly growing field of nanotechnology, in that new data can inform and modify research plans, funding, processes, and risk-management decisions. Review also is needed for basic-research inputs into the committee's strategy; for example, review and assessment of the availability and accessibility of and gaps in data on nanomaterials, products, and their uses.

Federal EHS research has been reviewed and updated periodically. In its *Strategy for Nanotechnology-Related Environmental, Health, and Safety Research (*NEHI 2008, p. 2), the NNI committed to reviewing and updating its plan "as research progresses." A year later, the 2008 NNI strategy was reviewed by the National Research Council (NRC 2009). In December 2010, the NNI released for public comment a draft *National Nanotechnology Initiative 2011 Environmental, Health, and Safety Strategy* (NEHI 2010), which updates, revises, and replaces its 2008 strategy. It states (pp. 2-3):

> Given the dynamic nature of research in this area, the NNI incorporated adaptive management into its first NNI EHS research strategy, the 2008 NNI Strategy for Nanotechnology-Related EHS Research to allow for modification of the strategy based on research progress, new findings, and product development. This document, the 2011 NNI EHS research strategy, is a result of that adaptive management process and revises and replaces the 2008 strategy. . . . The adaptive management process remains part of the 2011 NNI EHS research strategy to ensure proactive, science-based management of engineering nanomaterials (ENMs) into the future. Ongoing evaluation of research progress is conducted by the Nanotechnology Environmental Health Implications (NEHI) Working Group. . . . They will review and evaluate progress on an annual basis to ensure that the NNI EHS research strategy and activities keep pace with the rapid development of nanotechnology and evolving information on the effects of human and environmental exposure to nanomaterials.

Those efforts to review and revise the federal EHS-research strategy are commendable. The committee notes that NEHI (2010) incorporates some notable enhancements, including criteria for setting priorities for research on nanomaterials and nanotechnology-enabled products; identifying the need for reference materials, protocols, and standards for standardized measurements and the need for standardized nomenclature and terminology; a commitment to leverage public-private partnerships and explore mechanisms for interagency joint solicitation of research; giving high priority to international engagement; and detailing steps to foster and improve interagency coordination, communication, stakeholder interaction, and dissemination of knowledge and information.

Given that the federal EHS-research efforts are not the sole input into the present committee's strategy, the committee encourages further consideration and integration of international, industrial, and other nongovernment research efforts. Continuing efforts are also needed to identify and assess changing market conditions, including the nanomaterials used and the products made and changing regulatory conditions, in that these influence the research needs and priorities articulated in the committee's strategy.

Indicators for Evaluating Progress

The Near Term

The committee is tasked with delivering a second report 18 months after publication of this one. Issues to be addressed in the second report include an assessment of progress in understanding the EHS aspects of nanotechnology and the extent to which the short-term and long-term research priorities have been initiated or implemented. This 18-month timeframe is clearly too short to have substantial new research programs in place, let alone evaluate actual research outcomes. But the committee considers that the timeframe is sufficient to see progress in initiating research in each of the four high-priority categories identified in Chapter 5 and progress in developing the infrastructure, accountability, and coordination mechanisms discussed in this chapter. Progress in addressing those foundational elements will go a long way toward ensuring effective support and management of the research required to provide information for identifying, assessing, and effectively managing the potential EHS consequences of ENMs.

In its next report, the committee will evaluate the extent to which the research in each of the four high-priority categories identified in Chapter 5 has been initiated and the strategy implementation issues raised in the conclusions in this chapter have been addressed. (See Boxes 6-2 and 6-3 for summaries of indicators of research and implementation progress, respectively.) For examining initiation of research, the committee will simply ask whether the specific research-progress indicators are being addressed; little or no evaluation will be possible. With respect to the strategy-implementation issues, the committee will be particularly attentive to progress in establishing institutional arrangements and mechanisms that foster interagency interaction, collaboration, and accountability; developing and implementing mechanisms for stakeholder engagement; efforts to advance integration among sectors and institutions involved in EHS research, including public-private partnerships; structural changes that address conflicts of interest; and informatics and information-sharing.

BOX 6-2 Research-Progress Indicators

Adaptive research and knowledge infrastructure for accelerating research progress and providing rapid feedback to advance the research

- Extent of development of libraries of well-characterized nanomaterials, including those prevalent in commerce and reference and standard materials.
- Development of methods for detecting, characterizing, tracking, and monitoring nanomaterials and their transformations in relevant media.
- Development of methods to quantify effects of nanomaterials in experimental systems.
- Advancement of systems for sharing the results of research and fostering development of predictive models for nanomaterial behaviors.
- Extent of joining together of existing databases, including development of common informatics ontologies.

Quantifying and characterizing the origins of nanomaterial releases

Progress indicators will be related to the short-term activities identified in Chapter 5:

- Developing inventories of current and near-term production of nanomaterials.
- Developing inventories of intended use of nanomaterials and value-chain transfers.
- Identifying critical release points along the value chain.
- Identifying critical populations or systems exposed.
- Characterizing released materials and associated receptor environments.
- Modeling nanomaterial releases along the value chain.

Processes affecting both exposure and hazard

- Steps taken toward development of a knowledge infrastructure able to describe the diversity and dynamics of nanomaterials and their transformations in relevant biologic and environmental media.
- Progress toward developing instrumentation to measure key nanomaterial properties and changes in them in relevant biologic and environmental media.
- Initiation of interdisciplinary research that can relate native nanomaterial structures to transformations that occur in organisms and as a result of biologic processes.
- Extent of use of experimental research results in initial models for predicting nanomaterial behavior in complex biologic and environmental settings.

(Continued)

BOX 6-2 Continued

Nanomaterial interactions in complex systems ranging from subcellular systems to ecosystems

- Extent of initiation of studies that address heretofore underrepresented fields of research, such as those seeking to relate in vitro to in vivo observations, to predict ecosystem effects, or to examine effects on the endocrine or developmental systems.
- Steps toward development of models for exposure and potential effects along the ecologic food chain.
- Extent of refinement of a set of screening tools that reflect important characteristics or toxicity pathways of the complex systems described above.
- Extent of adaptation of existing system-level tools (such as individual species tests, microcosms, and organ-system models) to support studies of nanomaterials in such systems.
- Identification of benchmark or reference materials, both positive and negative, for use in such studies and measurement tools and methods to estimate exposure and dose in those complex systems.

The Longer Term

As discussed in Chapter 1, the National Research Council Committee on Research Priorities for Airborne Particulate Matter was charged with developing and monitoring progress in implementing a similar risk-research strategy. That committee developed six criteria that it used to evaluate progress in conducting the high-priority research and in implementing the strategy (NRC 1998, 1999):

- **Scientific value**: Does the research fill critical knowledge and data gaps?
- **Decision-making value:** Does the research reduce uncertainties and inform decision-making by key stakeholders, for example, decisions about risk assessment and risk management?
- **Feasibility and timing:** Is the research technically and economically feasible, and can it be done in a timeframe responsive to stakeholder and decision-maker needs?
- **Interaction and collaboration:** How well does the research agenda foster the collaboration and interaction needed among scientific disciplines, agencies, academe, and private sector, especially in addressing cross-cutting issues? Are the scientific expertise, capacity, and resources appropriately used to enhance scientific creativity, quality, and productivity?

BOX 6-3 Indicators of Progress in Implementation

Enhancing interagency coordination

- Progress toward establishing a mechanism to ensure sufficient management and budgetary authority to develop and implement an EHS-research strategy among NNI agencies.
- Extent to which NNCO is annually identifying funding needs for interagency collaboration on critical high-priority research.

Providing for stakeholder engagement in the research strategy

- Progress toward actively engaging diverse stakeholders in a continuing manner in all aspects of strategy development, implementation, and revision.

Conducting and communicating the results of research funded through public-private partnerships

- Progress toward establishment of effective public-private partnerships, as measured by such steps as completion of partnership agreements, issuance of requests for proposal, and establishment of a sound governance structure.

Managing potential conflicts of interest

- Progress toward achieving a clear separation in management and budgetary authority and accountability between the functions of developing and promoting applications of nanotechnology and understanding and assessing its potential health and environmental implications.
- Continued separate tracking and reporting of EHS research activities and funding distinct from those for other, more basic or application-oriented research.

- **Integration:** How well is the research agenda coordinated and integrated with respect to planning, budgeting, and management, including between government and private organizations?
- **Accessibility:** How well is information about research plans, budgets, progress, and results made accessible to agencies, research organizations, and interested stakeholders?

The present committee believes that the same criteria should be used to evaluate the extent of longer-term progress in implementing the research agenda proposed in this report. The criteria should be applied in evaluating research progress periodically. The committee notes that the NNI has already made substantial progress in addressing the latter three criteria (NEHI 2008, 2010).

As previously stated, the planning, management, and implementation of the research strategy are just as critical as the identification of the research priorities themselves. Without careful attention to those processes, any research agenda will fall short of expectations, no matter how compelling and well reasoned it may be.

To evaluate research progress later, the committee suggests applying specific longer-term progress indicators that correspond to the criteria presented in NRC (1998, 1999).

Scientific Value

The scientific value of the information generated can be assessed in terms of its overall contribution to enhancing understanding of the EHS effects of ENMs and reducing the uncertainty faced by stakeholders who must make decisions about nanotechnology and managing its potential risks. As noted in Chapter 5, such scientific knowledge will fill important data gaps and provide information on what the committee believes are the most critical elements and interactions for understanding EHS effects and determining whether a material is harmful. This includes knowledge about pathways and the likelihood of exposure through the life cycle and value chain, exposure of relevant targets, activation of pathways of disease and organism effects, and resulting effects on the health of humans and ecosystems. Specific progress indicators include

- Number, distribution, and adequacy of research projects that address priorities, gaps, and critical interactions in each part of the research agenda, including the number of research-agenda priorities planned and initiated, even if not completed, and the number of strategy-related research project applications received and funded.
- Distribution and adequacy of research through the life cycle and value chain of ENMs.
- Usefulness of study results for forming hypotheses for future research.
- Usefulness of research in building new research capacity, skills, and tools for future research.
- Extent to which uncertainty about human health and environmental risks is reduced.

Decision-Making Value

The decision-making value of the knowledge generated can be assessed in terms of its usefulness to the stakeholders who must make decisions about the development, production, and life-cycle use of ENMs. They include government agencies charged with protecting human and ecosystem health; developers, producers, suppliers, and purchasers of ENMs who must make economic and risk-management decisions in the face of scientific and regulatory uncertainty about EHS effects; and consumers of nanomaterial-enabled products. In addition to providing scientific information that reduces uncertainty about key decisions for stakeholders, the agenda should generate and make accessible basic information about the materials and products themselves—information critical for the research enterprise and important to many stakeholders. That includes information about the nanomaterials and products being produced, planned, or envisioned; identification of populations potentially exposed and at risk; prevention and control measures and practices in place or needed for protection and precaution in the face of uncertainty; identification of nanomaterial-enabled consumer products and information on consumer use; and information about disposal practices. Specific progress indicators include

- The extent to which the research has generated knowledge or information useful for decision-makers and other stakeholders. To what extent has it identified (and, ideally, reduced) the magnitude of uncertainty about EHS effects of ENMs? For example, to what extent has the new knowledge been integrated into risk-assessment decisions or regulatory processes? To what extent has it influenced private-sector research, investment, or production decisions related to nanomaterials?
- The extent to which the research has informed risk-management decisions—by government, industry, workers, and the public.
- The extent to which basic information about nanomaterials has become available and accessible to researchers, decision-makers, and the public—for example, information on nanomaterials and products containing nanomaterials currently produced and in use, data on exposure and exposure pathways, identification of populations at risk, preventive measures or practices in place or needed in the face of uncertainty, disposal practices, and consumer use.
- The usefulness of results in defining adverse effects on human health and ecosystems.
- The usefulness of results in identifying susceptible populations.

Timing and Feasibility

This indicator is related to the operational, technical, and economic feasibility of the research. Can it realistically be done in a timeframe responsive to decision-makers' needs? Specific progress indicators include

- Whether the technical tools and approaches to conduct the research are available or under development.
- Whether the research capacity and expertise for conducting the research are adequate and whether they are in the government, private sector, or academe.

- Whether the research is appropriately sequenced. Has a timeframe been articulated for each component of the research?
- Whether funding for the task is available and adequate.

Interaction and Collaboration

The scientific expertise, research capacity, and decision-making authorities needed to ensure the safety of ENMs are varied. They are found in a host of government, private-sector, and nonprofit organizations and are embedded in multiple disciplines from chemistry, biology, toxicology, medicine, and public health to engineering, computer modeling, and informatics. The committee believes that multidisciplinary interactions and collaborations both domestically and internationally are essential for progress in understanding and addressing the EHS dimensions of ENMs. Specific progress indicators include

- Specification of disciplines, expertise, and skills needed to achieve specific research objectives.
- Cooperative use of resources, including mutually funded or conducted research.
- Multidisciplinary collaboration in research projects.
- Joint workshops and conferences, and presentations and publications across disciplinary boundaries.
- Public-private research partnerships on specific elements of the EHS-research strategy.
- Sharing of databases and information among agencies and disciplines.
- Consistent use of terminology and measures among disciplines.
- Stakeholder engagement and participation in all aspects of the research agenda.
- Public participation in implementing the research strategy.

Integration

Given the agency-based government appropriations process, the different mandates and structures in and among government agencies, and the capacities and resources of private organizations, it is challenging to integrate the planning, budgeting, management, and monitoring of EHS research. But it is the very plethora of institutions, resources, and capacities that warrants efforts to integrate and optimize the use of federal resources, talent, and infrastructure in pursuit of the research agenda. Specific progress indicators include

- Collaborative or coordinated planning, monitoring, and evaluation of research both domestically and internationally, including multistakeholder engagement.

- Formal processes for exchanging and integrating knowledge, experience, and expertise of the EHS research communities in the implementation, monitoring, and evaluation of the research agenda.
- Appropriate balance and differentiation of kinds and loci of research activities, for example, applied vs basic, health vs ecosystem risk, intramural vs extramural, and public vs private.
- Use of a full spectrum of funding mechanisms.
- Mutually funded or conducted research.
- Data-sharing and model-sharing among investigators in and outside the federal government.

Accessibility

The ultimate outcome of EHS research is the prevention of harm or effective management of risks associated with ENMs. Thus, the knowledge and information generated must be accessible to and shared with scientists, research sponsors, decision-makers, the public, and others interested or with a role in risk prevention and management. Specific progress indicators include

- An accessible, searchable central database of research plans, studies, progress, funding, and results open to the research community, key stakeholders, and other interested parties.
- An "evergreen" database of nanomaterials, products, applications, and uses, and information on effective exposure-control technologies and practices.
- Outreach and provision of information to target, at-risk populations about EHS exposures, risks, risk prevention, and risk management.
- Periodic reports that summarize the status of research activities, synthesize research results and accomplishments, and identify remaining knowledge gaps.

Because the lifetime of the present committee is too short to provide for an assessment of research progress in understanding the EHS aspects of nanotechnology, the committee recommends that a rigorous and critical evaluation of the progress made in implementing and conducting research in keeping with its strategy be conducted within 5 years after completion of the committee's second report. That timeframe should be sufficient to observe progress and identify any needed changes in research directions and additional steps to maintain momentum toward addressing the identified high-priority research.

RESOURCES

There have been repeated expressions of concern that the federal funding devoted to EHS research on nanomaterials is insufficient (GAO 2008; Maynard

2008). That concern was echoed in the NRC (2009) review of the federal strategy (NEHI 2008). NRC (2009) also expressed concern that the federal strategy did not identify resources necessary to address questions concerning nanotechnology-related EHS-research needs. Specifically, there was no assessment of whether the aggregate level of spending by the federal agencies was adequate to address EHS-research needs or whether the resource expenditures by the agencies were appropriate to address EHS-research needs based on their own missions (p. 30). NRC (2009) recommended development of a strategy that included the "estimated resources that would be needed to address the [research] gap over a specified time frame."

In Chapter 5 of this report, the committee calls for maintenance of core EHS research funding at about $120 million per year over the next 5 years, as well as a strategic realignment of the federal resources being devoted to nanotechnology-related EHS research. The committee also recommends that modest additional resources from the private and public sectors, both nationally and internationally, augment the infrastructure needed to support an effective research program. The committee acknowledges, but has not attempted to estimate the resources (in addition to those required to conduct the research), needed for effective implementation of this strategy.

KEY AUDIENCES NEEDED TO IMPLEMENT THE STRATEGY

Implementation of the EHS-research strategy will require the coordinated participation of numerous entities—government, private sector, academic, and nongovernment organizations.

Government organizations have multiple roles related to understanding the potential implications of ENMs. They include conducting and supporting EHS research; ensuring coordination with ongoing research activities occurring internationally; responsibility for protecting workers, consumers, the general public, and the environment from adverse effects that may arise from exposure to nanomaterials in the workplace, products, and the general environment; and providing access to and assessing EHS-relevant information.

Private-sector nanomaterial developers and suppliers are core drivers and holders of nanotechnology-related EHS information. Manufacturers, nanomaterial suppliers, and their customers are the primary producers and handlers of the materials, so their input and knowledge are essential to the research agenda. For example, carbon nanotube (CNT) producer Bayer MaterialScience (BMS) conducts much of its toxicologic studies internally through Bayer Health Care. However, BMS is also involved in publicly supported studies through the project Nanotechnology Capacity Building NGOs, a European program to increase the understanding of nanomaterial-related EHS risks. And BMS—with other European CNT suppliers, such as Arkema and Nanocyl—is conducting workplace-exposure studies with the Producers Association of Carbon nanoTubes in Europe (PACTE) (Lux Research 2009).

Academic and research institutions also play a crucial role, especially in the fundamental research relevant to understanding EHS implications of nanotechnology. Academic researchers publish a large share of peer-reviewed articles on nanotechnology-related EHS research and provide expertise relevant to development, implementation, and evaluation of an effective research strategy.

Nongovernment and consumer organizations provide an additional perspective and expertise, a voice for the general public, and a valuable means of monitoring the overall efforts and progress of a research agenda. For example, they have highlighted the need for an accessible repository of EHS data to inform the public about the uses of and risks posed by nanomaterials (Lux Research 2009).

CONCLUDING REMARKS

The committee was charged with developing an integrated research strategy for addressing EHS aspects of ENMs. The committee recognizes that the success and impact of the proposed strategy depend on the institutional arrangements for its implementation and maintenance. This chapter has addressed critical issues related to coordination, collaboration, and leadership. The committee urges that these issues receive high priority because their resolution is integral to the success of the proposed research strategy.

REFERENCES

Callan, L.J. 1998. Organizational Conflict of Interest Regarding Department of Energy Laboratories. Memorandum to the Commissioners, from L. Joseph Callan, Executive Director for Operations, U.S. Nuclear Regulatory Commission. SECY-98-003. January 6, 1998 [online]. Available: http://www.nrc.gov/reading-rm/doc-collections/commission/secys/1998/secy1998-003/1998-003scy.html [accessed Mar. 17, 2011].

CENR (Committee on Environment and Natural Resources). 2011. Air Quality Research Subcommittee [online]. Available: http://www.esrl.noaa.gov/csd/aqrs/ [accessed May 16, 2011].

Denison, R. 2007a. Questions for the record to Dr. Richard A. Denison, U.S. House of Representatives: Research on Environmental and Safety Impacts of Nanotechnology: Current Status of Planning and Implementation under the National Nanotechnology Initiative, October 31, 2007 [online]. Available: http://www.edf.org/documents/7347_DenisonQFRresponsesFINAL.pdf [accessed Mar. 15, 2011].

Denison, R. 2007b. Statement of Richard D. Denison, Senior Scientist, Environmental Defense, before the House of Representatives Committee on Science At Hearing on Research on Environmental and Safety Impacts of Nanotechnology: Current Status Planning and Implementation under the National Nanotechnology Initiative, October 31, 2007 [online]. Available: http://www.edf.org/documents/7287_DenisonTestimony_10312007.pdf [accessed Mar. 15. 2011].

Dingbaum, S.D. 2002. Audit of NRC Oversight of ITS Federally Funded Research and Development Center (OIG-02-A-11). Memorandum Report to William D. Travers,

Executive Director for Operations, from Stephen D. Dingbaum, Assistant Inspector General for Audits, U.S. Nuclear Regulatory Commission. May 28, 2002 [online]. Available: www.nrc.gov/reading-rm/doc-collections/insp-gen/2002/02a-011.pdf [accessed Mar. 18, 2011].

DTSC/ICEMN (California Department of Toxic Substances Control/Industry Consortium for Environmental Measurement of Nanomaterials). 2011. DTSC and Industry Consortium for Environmental Measurement of Nanomaterials Workshop: Measurement Strategies for Nanomaterials in Media – Applicability to the Environment. June 2011 [online]. Available: http://www.dtsc.ca.gov/technologydevelopment/nanotechnology/nanoevents.cfm#June_2011 [accessed Oct. 5, 2011].

EPA (U.S. Environmental Protection Agency). 1987. P. 9 in U.S. EPA's Indoor Air Quality Implementation Plan. A Report to Congress under Title IV of the Superfund Amendments and Reauthorization Act of 1986: Radon Gas and Indoor Air Quality Research. EPA/600/8-87/031. Office of Air and Radiation, U.S. Environmental Protection Agency, Washington, DC.

EPA (U.S. Environmental Protection Agency). 1989. Pp. 34-35 in Report to Congress on Indoor Air Quality, Vol. Federal Programs Addressing Indoor Air Quality. EPA/400/1-89/001B. Office of Air and Radiation, Office of Research and Development, U.S. Environmental Protection Agency, Washington, DC.

Fenstermacher, D., C. Street, T. McSherry, V. Nayak, C. Overby, and M. Feldman. 2005. The Cancer Biomedical Informatics Grid (caBIG ™). Pp. 743-746 in Proceedings of the 2005 IEEE Engineering in Medicine and Biology Society 27th Annual Conference, September 1-4, 2005, Shanghai, China [online]. Available: http://ieeexplore.ieee.org/stamp/stamp.jsp?arnumber=1616521 [accessed June 9, 2011].

FNIH (Foundation for the National Institute of Health, Inc.). 2011. Foundation for the National Institute of Health, Inc. [online]. Available: http://www.fnih.org/ [accessed June 10, 2011].

GAO (U.S. General Accountability Office). 2008. Report to Congressional Requestors - Nanotechnology: Better Guidance is Needed to Ensure Accurate Reporting of Federal Research Focused on Environment, Health, and Safety Risks. GAO-08-402. Washington, DC: U.S. General Accountability Office. March 2008 [online]. Available: http://www.gao.gov/new.items/d08402.pdf [accessed May 9, 2011].

Gene Ontology. 2011. Welcome to the Gene Ontology website! [online]. Available: http://www.geneontology.org/ [accessed May 16, 2011].

HEI (Health Effects Institute). 2011. The Health Effects Institute [online]. Available: http://www.healtheffects.org [accessed Mar. 15, 2011].

iPlant Collaborative. 2011. iPlant Collaborative Empowering A New Plant Biology [online]. Available: http://www.iplantcollaborative.org/ [accessed June 8, 2011].

Lux Research. 2009. Nanotech's Evolving Environmental, Health, and Safety Landscape: The Regulations Are Coming. Lux Research. October 2, 2009 [online]. Available: https://portal.luxresearchinc.com/research/document_excerpt/5578 [accessed June 10, 2011].

Maynard, A. 2007. Testimony to Committee on Science and Technology, U.S. House of Representatives: Research on Environmental and Safety Impacts of Nanotechnology: Current Status of Planning and Implementation under the National Nanotechnology Initiative, October 31, 2007 [online]. Available: http://www.nanotechproject.org/process/files/5896/maynardtestimony_10_31_07.pdf [accessed Mar. 15, 2011].

Maynard, A.D. 2008. Testimony to Committee on Science and Technology, U.S. House of Representatives: The National Nanotechnology Initiative Amendments Act of

2008, Annex A. Assessment of U.S. Government Nanotechnology Environmental Safety and Health Research for 2006. April 16, 2008 [online]. Available: http://democrats.science.house.gov/Media/File/Commdocs/hearings/2008/Full/16apr/Maynard_Testimony.pdf [accessed May 9, 2011].

Nanoinformatics. 2010. Nanoinformatics 2010: A Collaborative Roadmapping and Planning Project, November 3-5, 2010, Arlington, VA [online]. Available: http://eprints.internano.org/606/1/Program-fullfinal.pdf [accessed June 5, 2011].

NAS/NAE/IOM (National Academy of Sciences, National Academy of Engineering, and Institute of Medicine). 2009. Ensuring the Integrity, Accessibility, and Stewardship of Research Data in the Digital Age. Washington, DC: National Academies Press.

Nanowerk News. 2010. National Nanotechnology Coordination Office Announces New Deputy Director. Nanowerk News, October 26, 2010 [online]. Available: http://www.nanowerk.com/news/newsid=18689.php [accessed Mar. 15, 2011].

NCI (National Cancer Institute). 2011a. Nanotechnology Characterization Laboratory, U.S. National Institutes of Health [online]. Available: http://ncl.cancer.gov/ [accessed Mar. 15, 2011].

NCI (National Cancer Institute). 2011b. caNanoLab [online]. Available: https://cananolab.nci.nih.gov/caNanoLab/ [accessed Dec. 14, 2011].

NEHI (Nanotechnology Environmental Health Implications Working Group). 2008. National Nanotechnology Initiative Strategy for Nanotechnology-Related Environmental, Health, and Safety Research. Arlington, VA: National Nanotechnology Coordination Office. February 2008 [online]. Available: http://www.nanowerk.com/nanotechnology/reports/reportpdf/report116.pdf [accessed June 10, 2011].

NEHI (Nanotechnology Environmental Health Implications Working Group). 2010. National Nanotechnology Initiative 2011 Environmental, Health, and Safety Strategy - Draft for Public Comment, December 2010. Nanotechnology Environmental Health Implications Working Group, Subcommittee on Nanoscale Science, Engineering, and Technology, Committee on Technology, National Science and Technology Council. [online]. Available: http://strategy.nano.gov/wp/ [accessed Feb. 14, 2011].

NEHI (Nanotechnology Environmental Health Implications Working Group). 2011a. National Nanotechnology Initiative 2011 Environmental, Health, and Safety Strategy, October 2011. Nanotechnology Environmental Health Implications Working Group, Subcommittee on Nanoscale Science, Engineering, and Technology, Committee on Technology, National Science and Technology Council [online]. Available: http://www.nano.gov/sites/default/files/pub_resource/nni_2011_ehs_research_strategy.pdf [accessed Nov. 23, 2011].

NEHI (Nanotechnology Environmental and Health Implications). 2011b. Nanotechnology Environmental and Health Implications (NEHI). Purpose and Goals. [online]. Available: http://www.nano.gov /NEHI [accessed June 17, 2011].

NIA (Nanotechnology Industries Association). 2011. Opportunity for Collaboration with California Department of Toxic Substances Control. February 8, 2011 [online]. Available: http://www.nanotechia.org/nia-internal-news/opportunity-for-collaboration-with-california-department-of-toxic-substances-control [accessed Oct. 5, 2011].

NNN (National Nanomanufacturing Network). 2011. Nanoinformatics 2020 Roadmap. Amherst, MA: National Nanomanufacturing Network. April 2011 [online]. Available: http://eprints.internano.org/607/1/Roadmap_FINAL041311.pdf [accessed May 5, 2011].

NORA (National Occupational Research Agenda). 2008. About NORA...Partnerships, Research and Practice. National Institute for Occupational Safety and Health, Centers for Disease Control and Prevention [online]. Available: http://www.cdc.gov/niosh/NORA/about.html [accessed Mar. 16, 2011].

NORA (National Occupational Research Agenda). 2011. NORA Cross-Sector Council. National Institute for Occupational Safety and Health, Centers for Disease Control and Prevention [online]. Available: http://www.cdc.gov/niosh/nora/councils/cross/default.html [accessed Mar. 16, 2011].

NRC (National Research Council). 1998. Research Priorities for Airborne Particulate Matter. I. Immediate Priorities and a Long-Range Research Portfolio. Washington, DC: National Academy Press.

NRC (National Research Council). 1999. Research Priorities for Airborne Particulate Matter. II. Evaluating Research Progress and Updating the Portfolio. Washington, DC: National Academy Press.

NRC (National Research Council). 2009. Review of the Federal Strategy for Nanotechnology-Related Environmental, Health, and Safety Research. Washington, DC: National Academies Press.

OSTP (Office of Science Technology and Policy). 2011. NSTC Committee on Environment, Natural Resources, and Sustainability. Office of Science Technology and Policy [online]. Available: http://www.whitehouse.gov/administration/eop/ostp/nstc/committees/cenrs [accessed May 16, 2011].

Teague, C. 2007. Testimony to Committee on Science and Technology, U.S. House of Representatives: Research on Environmental and Safety Impacts of Nanotechnology: Current Status of Planning and Implementation under the National Nanotechnology Initiative, October 31, 2007 [online]. Available: http://www.nano.gov/html/res/200711Teague_testiomony.pdf [accessed Mar. 15, 2011].

Thomas, D.G., R.V. Pappu, N.A. Baker. 2011. NanoParticle ontology for cancer nanotechnology research. J. Biomed. Inform. 44(1):59-74.

Travers, W.D. 1999. Organizational Conflict of Interest Regarding Department of Energy Laboratories. Memorandum to the Commissioners, from William D. Travers, Executive Director for Operations, U.S. Nuclear Regulatory Commission. SECY-99-043. February 5, 1999 [online]. Available: http://www.nrc.gov/reading-rm/doc-collections/commission/secys/1999/secy1999-043/1999-043scy.html [accessed Mar. 17, 2011].

U.S. NRC (U.S. Nuclear Regulatory Commission). 2010a. NRC-Regulator of Nuclear Safety. (NUREG/BR-0164, Rev. 4). U.S. Nuclear Regulatory Commission [online]. Available: http://www.nrc.gov/reading-rm/doc-collections/nuregs/brochures/br0164/r4/ [accessed Mar. 16, 2011].

U.S. NRC (U.S. Nuclear Regulatory Commission). 2010b. Our History. U.S. Nuclear Regulatory Commission [online]. Available: http://www.nrc.gov/about-nrc/history.html [accessed Mar. 16, 2011].

U.S. NRC (U.S. Nuclear Regulatory Commission). 2010c. Nuclear Regulatory Research. Office of Nuclear Regulatory Research, U.S. Nuclear Regulatory Commission [online]. Available: http://www.nrc.gov/about-nrc/organization/resfuncdesc.html [accessed Mar. 16, 2011].

Appendix A

Biographic Information on the Committee to Develop A Research Strategy for Environmental, Health, and Safety Aspects of Engineered Nanomaterials

Jonathan M. Samet (*Chair*) is a pulmonary physician and epidemiologist. He is a professor and Flora L. Thornton Chair of the Department of Preventive Medicine of the Keck School of Medicine of the University of Southern California (USC) and director of the USC Institute for Global Health. Dr. Samet's research has focused on the health risks posed by inhaled pollutants. He has served on numerous committees concerned with public health: the U.S. Environmental Protection Agency Science Advisory Board; committees of the National Research Council, including chairing the Committee on Biological Effects of Ionizing Radiation VI, the Committee on Research Priorities for Airborne Particulate Matter, and the Board on Environmental Studies and Toxicology; and committees of the Institute of Medicine (IOM). He is a member of IOM and of the Committee on Science, Technology, and Law. Dr. Samet received his MD from the University of Rochester School of Medicine and Dentistry.

Tina Bahadori is the National Program Director for Chemical Safety and Sustainability at the U.S. Environmental Protection Agency (EPA). Before joining EPA in May 2012, she was the managing director of the Long-Range Research Initiative program of the American Chemistry Council. Dr. Bahadori is the past president of the International Society of Exposure Science and is an associate editor of the *Journal of Exposure Science and Environmental Epidemiology*. She has served as a member of several committees of the National Academies, as a peer reviewer for EPA grants and programs, as a member of the Exposure to Chemical Agents Working Group for the National Children's Study, and as a

Study, and as a member of the Centers for Disease Control and Prevention National Center for Environmental Health-Agency for Toxic Substances and Disease Registry National Conversation on Public Health and Chemical Exposure Leadership Council. Before joining the ACC, she was the manager of Air Quality Health Integrated Programs of the Electric Power Research Institute. Dr. Bahadori holds a doctorate in environmental science and engineering from the Harvard School of Public Health.

Jurron Bradley joined BASF as a clean energy market manager in June 2011. In this role, he is responsible for creating BASF's first market facing-unit for the clean energy industry. Before joining BASF, Jurron led the consulting team at Lux Research, which provides clients with strategic advice on technology, including nanotechnology, and market trends and themes. Before joining Lux Research, Dr. Bradley worked at Praxair, Inc., where he designed air separation and argon recycling plants and managed a thermodynamics laboratory. He also led research efforts to reduce mercury emissions from coal-fired boilers and worked on the development of technology to reduce emissions of nitrogen oxides from coal-fired boilers. Dr. Bradley later joined Praxair's technology planning and strategy group in which he played a key role in developing strategic approaches for the entire research and development organization. Dr. Bradley received a PhD in chemical engineering from the Georgia Institute of Technology.

Seth Coe-Sullivan is a cofounder and chief technology officer of QD Vision. His work spans quantum dot materials; new fabrication techniques, including thin-film deposition equipment design; and device architectures for efficient QD-LED light emission. Dr. Coe-Sullivan has more than 20 papers and patents pending in the fields of organic light-emitting devices, quantum dot LEDs, and nanotechnology fabrication. He was awarded *Technology Review* magazine's TR35 Award in 2006 as one of the top 35 innovators under the age of 35 years. In 2007, *BusinessWeek* named him one of the top young entrepreneurs under the age of 30 years, and in 2009, he was a finalist for the Mass Technology Leadership Council's CTO of the year. Dr. Coe-Sullivan serves on Brown University's Engineering Advisory Council. He received his PhD in electrical engineering from the Massachusetts Institute of Technology; his thesis work on incorporating quantum dots into hybrid organic-inorganic LED structures led to the formation of QD Vision.

Vicki L. Colvin is vice provost for research, professor of chemistry, and director of the Center for Biological and Environmental Nanotechnology (CBEN) at Rice University. Among CBEN's primary interests is the application of nanotechnology to the environment. She has received numerous accolades for her teaching abilities, including Phi Beta Kappa's Teaching Prize for 1998-1999 and the Camille Dreyfus Teacher Scholar Award in 2002. In 2002, she was also named one of *Discover* magazine's Top 20 Scientists to Watch and received an

Alfred P. Sloan Fellowship. In 2007, she was named a fellow of the American Association for the Advancement of Science (AAAS). Dr. Colvin is a frequent contributor to *Advanced Materials*, *Physical Review Letters*, and other peer-reviewed journals and holds patents to seven inventions. Dr. Colvin served on the NRC Committee for Review of the Federal Strategy to Address Environmental, Health, and Safety Research Needs for Engineered Nanoscale Materials. She received her PhD in chemistry from the University of California, Berkeley, where she was awarded the American Chemical Society's Victor K. LaMer Award for her work in colloid and surface chemistry.

Edward D. Crandall is the Hastings Professor and Kenneth T. Norris, Jr. Chair of Medicine and chair of the Division of Pulmonary and Critical Care Medicine of the Keck School of Medicine of the University of Southern California. Dr. Crandall's clinical interests include critical-care medicine and pulmonary disease. He has written numerous peer-reviewed articles on cardiopulmonary biology. His specific research interests are in the regulation of the differentiation and transport properties of alveolar epithelial cells. He is actively involved in research on the interactions of nanomaterials with alveolar epithelium. Dr. Crandall received his PhD from Northwestern University and his MD from the University of Pennsylvania.

Richard A. Denison is a senior scientist at the Environmental Defense Fund. Dr. Denison has 27 years of experience in the environmental arena, specializing in chemical policy and hazards, exposure, and risk assessment and management of industrial chemicals and nanomaterials. He is a member of the National Research Council Board on Environmental Studies and Toxicology and serves on the Green Ribbon Science Panel for California's Green Chemistry Initiative. Dr. Denison was a member of the National Pollution Prevention and Toxics Advisory Committee, which advised the Environmental Protection Agency's Office of Pollution Prevention and Toxics. Previously, Dr. Denison was an analyst and assistant project director in the Oceans and Environment Program of the Office of Technology Assessment of the U.S. Congress. Dr. Denison received his PhD in molecular biophysics and biochemistry from Yale University.

William H. Farland is the senior vice president for research of Colorado State University and a professor in its Department of Environmental and Radiological Health Sciences in the School of Veterinary Medicine and Biomedical Sciences. In 2006, Dr. Farland was appointed deputy assistant administrator for science in the Environmental Protection Agency (EPA) Office of Research and Development (ORD). He had served as the acting deputy assistant administrator since 2001. In 2003, Dr. Farland has also been chief scientist in the Office of the Agency Science Adviser. He served as EPA's acting science adviser throughout 2005. Formerly, he was the director of ORD National Center for Environmental Assessment. Dr. Farland served on a number of executive-level committees and advisory boards in the federal government. In 2005-2006, he chaired the Executive Committee of the National Toxicology Program. He was also a member of the Scientific Advisory Council of the Risk Sciences and Public Policy Institute of the Johns Hopkins University School of Hygiene and Public Health, a public member of the American Chemistry Council's Strategic Science Team for its Long-Range Research Initiative, and a member of the Programme Advisory Committee for the World Health Organization's International Programme on Chemical Safety. Dr. Farland recently served as chair of an external advisory group for the National Institute of Environmental Health Sciences regarding the future of the Superfund Basic Research Program. He is the chair of a standing Committee on Emerging Science for Environmental Health Decisions of the National Research Council. In 2002, Dr. Farland was recognized by the Society for Risk Analysis with the Outstanding Risk Practitioner Award, and in 2005, he was named a fellow of the society. In 2006, he received a Presidential Rank Award for his service as a federal senior executive. In 2007, he was elected a fellow of the Academy of Toxicological Sciences. Dr. Farland received his PhD from the University of California, Los Angeles in cell biology and biochemistry.

Martin Fritts is a senior principal scientist who supported the Nanotechnology Characterization Laboratory and SAIC-Frederick in accelerating the transition of nanotechnology to cancer and biomedical applications. He is also a computational and experimental physicist who works on the implementation of advanced imaging and measurement instrumentation, modeling, and simulation to elucidate the structure-activity relationships of nanomaterials and informatics systems to advance knowledge-sharing. Dr. Fritts serves as the cochair of the American Society for Testing and Materials E56.02 Subcommittee on Nanotechnology Characterization. Before joining SAIC-Frederick, he developed and prototyped nanotechnology applications for industry and government through SAIC's Nanotechnology Initiatives Division. He earned a PhD in nuclear physics from Yale University.

Philip K. Hopke is the Bayard D. Clarkson Distinguished Professor in the Department of Chemical and Biomolecular Engineering and the Department of Chemistry of Clarkson University. He is also director of the university's Center for the Environment and its Center for Air Resources Engineering and Sciences. His research interests are related primarily to particles in the air, including particle formation, sampling and analysis, composition, and origin. His current projects are related to receptor modeling, ambient monitoring, and nucleation. Dr. Hopke has been elected to membership in the International Statistics Institute and is a fellow of the American Association for the Advancement of Science. He is also a fellow of the American Association for Aerosol Research, in which he has served in various roles, including president, vice president, and member of the board of directors. Dr. Hopke is a member of the American Institute of Chemical Engineers, the International Society of Exposure Science, and the International Society of Indoor Air Quality and Climate, and others. He has served as a member of the U.S. Environmental Protection Agency Advisory Council on

Clean Air Act Compliance Analysis and as a member of several National Research Council committees, most recently the Committee on Energy Futures and Air Pollution in Urban China and the United States, the Committee on Research Priorities for Airborne Particulate Matter, and the Committee on Air Quality Management in the United States. Dr. Hopke received his PhD in chemistry from Princeton University.

James E. Hutchison is the Lokey-Harrington Professor of Chemistry at the University of Oregon. He is the founding director of the Oregon Nanoscience and Microtechnologies Institute for Safer Nanomaterials and Nanomanufacturing Initiative, a virtual center that unites 30 principal investigators in the Northwest around the goals of designing greener nanomaterials and nanomanufacturing. Dr. Hutchison's research focuses on molecular-level design and synthesis of functional surface coatings and nanomaterials for a wide array of applications, in which the design of new processes and materials draws heavily on the principles of green chemistry. Dr. Hutchison received several awards and honors, including the Alfred P. Sloan Research Fellowship and the National Science Foundation CAREER Award. He was a member of the National Research Council Committee on Grand Challenges for Sustainability in the Chemistry Industry. Dr. Hutchison received his PhD in organic chemistry from Stanford University.

Rebecca D. Klaper is an associate professor in the School of Freshwater Sciences, University of Wisconsin-Milwaukee. The School of Freshwater Sciences (at the Great Lakes WATER Institute) is dedicated to providing basic and applied research to inform policy decisions involving freshwater resources. Dr. Klaper uses traditional toxicologic methods and genomic technologies to study the potential effects of emerging contaminants, such as nanoparticles and pharmaceuticals, on aquatic organisms. Dr. Klaper received an American Association for the Advancement of Science Science and Technology Policy Fellowship, in which she worked in the National Center for Environmental Assessment at the Environmental Protection Agency (EPA). She has served as an invited scientific expert to both the U.S. National Nanotechnology Initiative and the Organisation for Economic and Co-operative Development Panel on Nanotechnology, for which she has testified on the potential effects of nanoparticles on the environment and the utility of current testing strategies. She has served as a technical expert in reviewing the EPA white paper on the environmental effects of nanotechnologies and the EPA research strategy for nanotechnology. She also was involved in writing the EPA white paper on the use of genomic technologies in risk assessment. Dr. Klaper received her PhD in ecology from the Institute of Ecology of the University of Georgia.

Gregory V. Lowry is a professor in the Department of Civil and Environmental Engineering of Carnegie Mellon University and deputy director of the National Science Foundation Center for Environmental Implications of Nanotechnology. He researches sustainable development of nanomaterials and nanotechnologies, including the fate, mobility, and toxicity of nanomaterials in the environment, remediation and treatment technologies that use nanomaterials, and nanoparticle-contaminant and biota interactions. He also works on sustainable energy via carbon capture and storage. His current projects include elucidating the role of adsorbed macromolecules on nanoparticle transport and fate in the environment, in situ sediment management with innovative sediment caps, dense nonaqueous-phase liquid source zone remediation through delivery of reactive nanoparticles to the nonaqueous-phase-water interface, and carbon dioxide capture, sequestration, and monitoring. Dr. Lowry served as an external advisory board member for the Center for Biological and Environmental Nanotechnology. He was a review panelist for the Environmental Protection Agency draft nanomaterial research strategy. He is a member of the American Chemical Society, the American Society of Civil Engineers, and the Association of Environmental Engineering and Science Professors. He received his PhD in civil-environmental engineering from Stanford University.

Andrew D. Maynard is the director of the Risk Science Center of the University of Michigan School of Public Health. He previously served as the chief science adviser in the Woodrow Wilson International Center for Scholars for the Project on Emerging Nanotechnologies. Dr. Maynard's research interests revolve around aerosol characterization, the implications of nanotechnology for human health and the environment, and managing the challenges and opportunities of emerging technologies. Dr. Maynard's expertise covers many facets of risk science, emerging technologies, science policy, and communication. Previously, he worked for the National Institute for Occupational Safety and Health and represented the agency on the Nanomaterial Science, Engineering and Technology (NSET) subcommittee of the National Science and Technology Council and cochaired the Nanotechnology Health and Environment Implications working group of NSET. He serves on the World Economic Forum Global Agenda Council on Emerging Technologies and is a member of the Executive Committee of the International Council on Nanotechnology. He previously chaired the International Standards Organization Working Group on size-selective sampling in the workplace. Dr. Maynard served as a member of the NRC Committee for Review of the Federal Strategy to Address Environmental, Health, and Safety Research Needs for Engineered Nanoscale Materials. He earned his PhD in aerosol physics from the Cavendish Laboratory of the University of Cambridge, UK.

Günter Oberdörster is a professor in the Department of Environmental Medicine of the University of Rochester, director of the University of Rochester Ultrafine Particle Center, principal investigator of a Multidisciplinary Research Initiative in Nanotoxicology, and head of the Pulmonary Core of the National Institute of Environmental Health Sciences Center Grant. His research includes the effects and underlying mechanisms of lung injury induced by inhaled nonfibrous and fibrous particles, including extrapolation modeling and risk assessment. His studies with ultrafine particles influenced the field of inhalation toxicology, raising awareness of the unique biokinetics and toxic potential of nano-sized particles. He has served on many national and international committees and is the recipient of several scientific awards. Dr. Oberdörster has served on several National Research Council committees, including the Committee on Research Priorities for Airborne Particulate Matter and the Committee on the Review of the Federal Strategy to Address Environmental, Health, and Safety Research Needs for Engineered Nanoscale Materials. He is on the editorial boards of the *Journal of Aerosol Medicine, Particle and Fibre Toxicology, Nanotoxicology,* and the *International Journal of Hygiene and Environmental Health* and is associate editor of *Inhalation Toxicology* and *Environmental Health Perspectives.* He earned his DVM and PhD (in pharmacology) from the University of Giessen, Germany.

Kathleen M. Rest is the executive director of the Union of Concerned Scientists (UCS), a science-based nonprofit. She manages the organization's day-to-day affairs, supervising programs on issues ranging from climate change and clean energy to global security. Dr. Rest came to UCS from the National Institute for Occupational Safety and Health (NIOSH) in the Centers for Disease Control and Prevention, where she was the deputy director for programs. Throughout her tenure at NIOSH, she held several leadership positions, including serving as the institute's acting director during the period of September 11, 2001, and the anthrax events that followed. Before her federal service, Dr. Rest served on the faculty of several medical schools—most recently as an associate professor in the Department of Family and Community Medicine of the University of Massachusetts Medical Center and an adjunct associate professor in the University of Massachusetts School of Public Health—where she taught occupational, environmental, and public health. She has extensive experience as a researcher and adviser on occupational and environmental health issues in various countries, such as the Netherlands, Slovakia, Poland, Romania, Canada, and Greece. Dr. Rest was a founding member of the Association of Occupational and Environmental Clinics, a national nonprofit organization committed to improving the practice of occupational and environmental health through information-sharing and collaborative research. She also served as the chairperson of the National Advisory Committee on Occupational Safety and Health. Dr. Rest earned her PhD in health policy from Boston University.

Mark J. Utell is a professor of medicine and environmental medicine, a director of occupational and environmental medicine, and former director of pulmonary and critical-care Medicine in the University of Rochester Medical Center. He serves as associate chairman of the Department of Environmental Medicine. His research interests have centered on the effects of environmental toxicants on the human respiratory tract. Dr. Utell has published extensively on the health effects of inhaled gases, particles, and fibers in the workplace and other indoor and outdoor environments. He is the co-principal investigator of an Environmental Protection Agency (EPA) Particulate Matter Center and chair of the Health Effects Institute's Research Committee. He has served as chair of EPA's Environmental Health Committee and on the Executive Committee of the EPA Science Advisory Board. He is a former recipient of the National Institute of Environmental Health Sciences Academic Award in Environmental and Occupational Medicine. Dr. Utell is currently a member of the National Research Council's Board on Environmental Studies and Toxicology. He previously served on the National Research Council Committee on Research Priorities for Airborne Particulate Matter, the Institute of Medicine (IOM) Committee to Review the Health Consequences of Service during the Persian Gulf War, and the IOM Committee on Biodefense Analysis and Countermeasures. He received his MD from Tufts University School of Medicine.

David B. Warheit received his PhD in physiology from Wayne State University School of Medicine in Detroit. Later, he received a National Institutes of Health (NIH) postdoctoral fellowship, and 2 years later, a Parker Francis Pulmonary Fellowship, both of which he took to the National Institute of Environmental Health Sciences to study mechanisms of asbestos-related lung disease with Arnold Brody. In 1984, he moved to the DuPont Haskell Laboratory to develop a pulmonary-toxicology research laboratory. His major research interests are pulmonary toxicity mechanisms and corresponding risks related to inhaled particles, fibers, and nanomaterials. He is the author or coauthor of more than 100 publications and has been the recipient of the International Life Sciences Institute (ILSI) Kenneth Morgareidge Award (1993, Hannover, Germany) for contributions in toxicology by a young investigator and the Robert A. Scala Award and Lectureship in Toxicology (2000). He has also attained diplomate status of the Academy of Toxicological Sciences (2000) and the American Board of Toxicology (1988). He has served on NIH review committees (NIH Small Business Innovation Research and NIH Bioengineering) and has participated in working groups of the International Agency for Research on Cancer, the European Centre for Ecotoxicology and Toxicology of Chemicals, Organisation for Economic Co-operation and Development, the ILSI Risk Science Institute, the ILSI Health and Environmental Sciences Institute, and the National Research Council. He has served on several journal editorial boards, including *Inhalation Toxicology and Toxicological Sciences* (as the current associate editor), *Particle and Fibre Toxicology*, *Toxicology Letters,* and *Nano Letters*. He is the chairman of the European Centre for Ecotoxicology and Toxicology of Chemicals Task Force on Health and Environmental Safety of Nanomaterials, serves on the National Institute for Occupational Safety and Health Board of Scientific Counselors, and is interim vice-president of the Nanotoxicology Specialty Section.

Mark R. Wiesner serves as director of the Center for the Environmental Implications of Nanotechnology, headquartered at Duke University, where he holds the James L. Meriam Chair in Civil and Environmental Engineering with appointments in the Pratt School of Engineering and the Nicholas School of Environment. Dr. Wiesner's research has focused on the applications of emerging nanomaterials to membrane science and water treatment and an examination of the fate, transport, and effects of nanomaterials in the environment. He was co-editor and author of *Environmental Nanotechnologies* and serves as associate editor of the journals *Nanotoxicology* and *Environmental Engineering Science.* Before joining the Duke University faculty in 2006, Dr.Wiesner was a member of the Rice University faculty for 18 years, where he held appointments in the Department of Civil and Environmental Engineering and the Department of Chemical Engineering and served as associate dean of engineering and director of the Environmental and Energy Systems Institute. Before working in academe, Dr. Wiesner was a research engineer with the French company Lyonnaise des Eaux, in Le Pecq, France, and a principal engineer with the environmental engineering consulting firm of Malcolm Pirnie, Inc., White Plains, NY. He received the1995 Rudolf Hering Medal from the American Society of Civil Engineers, of which he is a fellow, and the 2004 Frontiers in Research Award from the Association of Environmental Engineering and Science Professors, on whose board he serves. In 2004, Dr. Wiesner was also named a de Fermat Laureate and was awarded an International Chair of Excellence in the Chemical Engineering Laboratory of the French Polytechnic Institute and National Institute for Applied Sciences in Toulouse, France. He received his PhD in environmental engineering from the Johns Hopkins University.

Appendix B

Implementation Scenarios: Informatics and Information-Sharing

Chapter 6 discussed the implementation of a fully integrated research strategy. Implementation of the strategy requires an infrastructure capable of expanding current institutional arrangements for interagency coordination, stakeholder engagement, public-private partnerships, and management of potential conflicts of interest. Implementation also requires new mechanisms for integrating informatics and information-sharing into the research structure. This appendix clarifies how a systems approach to the planning and development of a facile and agile informatics infrastructure might help to implement a research strategy that is responsive to the input of stakeholders and accelerate the research for nanotechnology applications and implications. This infrastructure would borrow heavily from advances in digital technologies and the Semantic Web to reduce collaboration timescales and ensure more effective communication among stakeholders.

Implementation scenarios will be summarized here for the development of methods and protocols, predictive and risk models, a federated data-sharing network, and a semantic informatics infrastructure. There are a great many similarities between challenges in improving model development and challenges in improving method development. Both require a systems overview of the problem to encompass the needs of the entire community, both need continual iterative development of pilot efforts to elicit user requirements and enlist user support, and both have an interest in implementing advanced digital tools and applications. The intent of the scenarios is to illustrate how a system approach could accelerate progress in nanoscience and nanotechnology research and translation; however, they should not be viewed as blueprints for implementation. Involving stakeholders at the beginning and throughout the development process is critical for successful implementation. The needs and requirements of the user base must be satisfied if best practices are to be advanced and critical participation in the informatics effort is to be attained.

In all the scenarios presented below, input from existing and emerging stakeholder and user communities is needed. Standard development organizations (SDOs) such as the International Organization for Standardization and the American Society for Testing and Materials (ASTM) continue to adopt new experimental protocols, guides, and practices from metrology institutes such as the National Institute of Standards and Technology, laboratories such as the National Cancer Institute's (NCI) Nanotechnology Characterization Laboratory, and industry. Interlaboratory testing of those protocols has been performed by the Asia-Pacific Economic Cooperation (Wang et al. 2007), the International Alliance for NanoEHS Harmonization (IANH 2011), and ASTM, among others, using nanomaterials recommended or developed by the Organisation for Economic Co-operation and Development, metrology institutes, or purchased commercially. Support for nanomaterial registration is becoming established through the National Institute of Biomedical Imaging and Bioengineering Registry. The stakeholder and user groups for modeling, data-sharing, and informatics infrastructure also are broadly based (Nanoinformatics 2011a). Historically, this larger nanotechnology community has interacted through workshops designed to support and harmonize increased collaboration and has adopted successful informatics implementations from other fields. More recently a series of workshops on nanoinformatics has provided a roadmap for collaborations in informatics (InterNano 2011) and to support the development and review of pilot nanoinformatics applications (Nanoinformatics 2011b), including those relevant to the following scenarios.

Method Development and Validation Scenario

This scenario closely follows international recommendations for standard method development and validation and adds video technologies to accelerate the development of the methods and for training. Key principles underlying this approach are that it should be efficient, be flexible, add value, be amenable to establishing data rights, provide for continual improvement of the protocol, document experience in its use, and be tailored to develop and maintain the entire needed dataset. A benefit of the scenario is that it provides a basis for training in new and revised methods and for accrediting contract research organizations (CROs) to allow more outsourcing of extensive nanomaterial characterization with validated methods. Finally, the scenario guarantees publication of sensitivity data that are not normally published but that are useful for establishing quantitative structure-activity relationships (QSARs), for designing and redesigning products and processes, and for assessing risk.

A possible set of best practices for establishing validated analytic methods on the basis of current practice and adapting them for nanomaterials to address the needs discussed in Chapter 4 includes the following:

1) Encourage early practitioners to document and publish protocols as individual methods are developed.

2) Begin collaborative development of new methods and protocols only after preliminary video protocols and sufficient well-characterized nanomaterials are available to support method development and testing.

3) Use a material registry system to designate unique lot descriptors for each nanomaterial sample, to maintain a catalog of descriptors to capture lot-to-lot variability in engineered nanomaterials (ENMs), and to correlate possibly different effects seen in various uses and analyses of material lots. Monitor shipping and record environmental conditions during transit of the nanomaterial and any biologic materials needed for use in interlaboratory studies (ILSs), including calibration efforts.

4) Accelerate progress in developing standard analytic methods by using video (particularly common video equipment and applications, such as cellular telephones) and digital collaboration environments (for example, wikis, RSS feeds, and Facebook) to facilitate broad participation and communication.

5) Conduct informal interlaboratory testing of the protocols to identify causes of laboratory bias and to investigate the ruggedness and reliability of the methods before development of the documentary standard. It may be faster to achieve consensus on the documentary standard because of the prior vetting of the protocol and the existence of the informal ILS results, robustness data, and video (JoVE 2011).

6) Rapidly modify the video protocol through a small ILS testing group so that consensus is reached and the (informal) error and uncertainty of the method are satisfactory.

7) Collect data to substantiate that the ruggedness and robustness of the method are adequate (quantification of the sensitivity of the method to variation in any experimental procedure, materials, or conditions is archived with the video protocol).

8) Develop a consensus documentary standard based on the video and the results of the informal ILSs and a more polished video illustrating the method.

9) Use the documentary standard and supplementary video in formal ILSs to determine the error and uncertainty of the method.

10) Publish

a) The documentary consensus standard.
b) The final video (to be used for training).
c) The error and uncertainty data and analysis.
d) The sensitivity data and analysis.

11) Establish a reporting standard for the level of validation achieved by a laboratory that is using the method (for example, full validation, partial validation, corroboration with at least one other laboratory, or a single laboratory result).

12) Continue to engage the stakeholder group through the collaborative environment to

a) Aggregate and organize information on experience with the method, especially details of sample preparation and additional controls required for testing of other ENMs.
b) Rapidly update the method, video, and databases.
c) Aid in establishing the minimum required characterization for each ENM.
d) Provide a virtual helpdesk for the method.

This scenario provides an overarching context for the different activities involved in developing a standard and a framework for collaboration among the various partners in the effort, such as standard development organizations (SDOs), metrology institutes, national user facilities, federally funded research and development centers, and CROs. The information infrastructure allows coordination among all participants in the collaboration, and although each participant may continue to perform its usual role, expansion of roles to new activities and products is possible.

Model Development and Validation Scenario

The informatics needs in Chapter 4 specified increasing the pace of model development and validation by leveraging digital technologies and applications; creating incentives for collaborative development and validation of models to estimate model error, uncertainty, and sensitivity more efficiently; providing platforms for continued improvement and testing of models; and archiving and curating model results. The needs assessment specified both specific capabilities required for an acceptable informatics infrastructure and those related to establishing new collaborative mechanisms. Model development has a rich history in the development and validation of models for biochemistry, pharmaceuticals, and genomics to delineate feasible recommendations for implementation. Model development and validation are different from method development and validation in that most models are already in a digital form and can readily be shared electronically. The scenario presented below poses a series of activities to provide common environments for interdisciplinary model building and validation for all the stakeholders involved in nanoscience and nanotechnology research and translation.

Nanomaterials require a variety of structural descriptors. For small nanoparticles such as dendrimers an exact structural description may be obtained as with polymers or proteins, even though the particle's conformation may be time dependent and change with its local environment and its interaction with other molecules and surfaces. For nanoparticles within the range of ultrafine particles, there may be large polydispersity and polymorphism, as these particles are primarily produced through batch synthesis rather than self assembly. The descriptors at this particle size range are primarily used in material science and include cruder measures such as core or shell sizes and ranges, as well as grain size and lattice defects. For both types of nanoparticles, modification of the surfaces of the particles contributes to polydispersity and interface chemistry and descriptors become increasingly important. Finally, the nanoparticles may be embedded in a matrix to fabricate a nanomaterial or nanoproduct (for example, colloids), resulting in a material with even greater polydispersity and thus a larger number of required descriptors.

Because of the variation in the properties of nanomaterials, understanding the mechanisms underlying these properties requires input from many different scientific and engineering fields. To develop a database for a particular nanomaterial, it is useful to have a large lot of material from the same manufactured batch, so that all disciplines contributing to the database are using the same material in their studies. Use of standard analytic methods is needed to generate high quality data and to determine the sensitivity of the experimentally measured values to changes in the material's environment. Similarly, modeling experiments should use a representative range of nanomaterial structures and validated, documented, and predictive models that can produce reliable results. The material repositories and registries, and the validated experimental protocols discussed previously have their analogs in modeling, but models have the advantage in that they can be more easily shared. The scenario below describes means of improving the quality and reliability of data derived from structural, predictive, and risk models.

Developing reliable structure-property relationships and their underlying physical, chemical, and biologic mechanisms for nanomaterials requires joint, informed, experimental modeling activities across many disciplines. Although it is difficult to determine experimentally the mechanisms for the interactions of nanomaterials with their biologic environments using polydisperse and polymorphic materials, a modeling effort that applies representative structures and validated models may provide important insights.

1) Develop key structural descriptors that provide basic definition of nanomaterial dimensions (and their dispersities), compositions, and surface chemistries. Although it may be possible to define only small nanoparticles precisely, descriptions for the purposes of informatics efforts must be defined as thoroughly as possible.

2) For selected existing applications, develop, archive, organize, curate, and validate molecular models for ENMs, including their surface coatings (and possibly weathered or transformed coatings). One existing pilot project is the Collaboratory for Structural Nanobiology (CSN 2011), which is modeled on the Worldwide Protein Data Bank (wwPDB 2011), and contains models for several classes of nanomaterials, viewing applications, model-building tools, and a wiki environment to facilitate collaborative activities.

3) Develop structural models and validate them by using blinded data on newly determined ENM structures. Compare the model's functionality with that of similar structures, and capture comment on the model's functionality and accuracy to develop requirements for an improved model.

4) Establish a user group (that represents all stakeholders) to draft a governance document for the structural model portal, taking advantage of the history and current governance of the wwPDB, especially regarding data rights, security, and the formation of a scientific advisory group for the next-generation portal.

5) Begin attempts to formulate classification schemes based on ENM "types," structural motifs, or correlations between molecular models and descriptors used in formulating QSARs for ENMs. This approach extends model development and validation efforts to predictive models, risk models, and functional models for biologic environments. A repository for similar models establishes a new means of collaborating on model development and of accelerating model development while ensuring that credit is assigned for researchers involved in developing and improving the models.

6) Use an initial pilot to archive, organize, curate, validate, and share predictive and probabilistic models and to set user requirements for a collaborative environment for developing and validating predictive models and submodels; organelle, cell, tissue, organ, system, organism, and ecosystem models; and probabilistic and risk models and submodels—with their associated files, run-time parameters, and test suites.

7) Collect the data taken to substantiate that the ruggedness and robustness of the models are adequate (this involves quantification of the sensitivity of the models to variation in any model parameters, materials, or conditions).

8) Establish a user group and a governance document for the portal for predictive model validation, especially with regard to data rights, security, and formation of a scientific advisory group.

9) Use the collaborative environment to create a readily accessible portal to aggregate and organize information on experience with the models tested to rapidly update the models, and to provide a virtual helpdesk for use of the models.

10) Once a model is adequately validated and has a large enough user group, port a copy of the model to a facility, like NanoHUB (NanoHUB.org 2011), to place an optimized version of the model, sample run-time parameters, and associated files and test suites for use in the larger user community. If computational models are stored in a repository with all the applications, programs, data files, scripts, and run-time parameters needed to replicate the original results, a common means of collaborating on model development is created, and the timescales to establish effective collaborations are reduced from years (in print media) to days.

Models of biologic systems—including organelle, cell, tissue, and organ—are used for many applications but are usually developed and validated in isolated efforts with little public record of either the process or the reference data. Such models could benefit from efforts to accelerate their development, validation, and reuse. Increased collaboration in QSAR and quantitative structure-property relationship (QSPR) development would provide a more complete picture of their accuracy and the correlation of their results with genomic, proteomic, and metabolomic data. Finally, there is a possible benefit in providing tools for collaborative code development for risk modeling and management. Although regulatory agencies have highly developed methods and processes for their own needs that are referenced to specific data streams, cross-fertilization may provide more rapid incorporation of novel computational methods, submodels, and techniques (Haimes 2009; IOM 2011).

A Scenario for Nomenclature and Terminology

Development of functional and adaptable nomenclature and terminologies for ENMs is critical for all informatics efforts. Adaptability is crucial because the precision with which ENM structure and composition are known is continuously improving. There have been many attempts to develop namespace terminologies by various organizations, such as SDOs. In many cases, the high-level concept definitions are inconsistent because the experts developing the terminology defined terms relevant to specific sub-disciplines, applications, or interest groups. As a result, concept definitions for a given term may differ substantially within a single SDO because sufficient resources are rarely available to achieve consistency or to produce properly framed definitions. Developing relationships among terms through simple taxonomies or ontologies also introduces a high degree of variability because different namespaces will set priorities for those relationships differently. That disparity has led to the development of mapping tools to aid in generating consistent mapping among terminologies, taxonomies, and ontologies within and among namespaces.

This scenario outlines steps that may be taken to accelerate the use of ontologies in achieving Semantic Web implementation for nanotechnology. It draws on the existing body of ontologies and expertise in semantic-tool development to provide a common infrastructure for semantic search to allow interoperable searching among databases curated by different scientific disciplines that return only what is specifically requested. The approach could promote common terminology but recognizes that different disciplines may use substantially different definitions for a given term or may impart nuance to such differences in ways that should be captured. That is, a semantic search capability recognizes that dictionaries may have a number of definitions for a given term, each relevant for a different namespace, and allows retrieval of all data relevant to the namespaces being searched. As in the previous scenarios, the relevance of this capability to implementation lies in the ability to query a federated system of databases, permitting transparent access into each discipline's data while using the terminology of the requester's discipline.

1) Solidify and continuously advance the precision of structural descriptors (for example, composition, size, shape, and surface coating) for ENMs to establish a basis for a functional nomenclature for these materials.

2) Tap existing pilot efforts in nanotechnology nomenclature, concept definitions, vocabulary, metadata, and ontologies to coalesce a Semantic Web (W3C 2011a) community of interest among stakeholders in nanoinformatics that spans several distinct disciplines. Examples include the NanoParticle Ontology for nanomedicine, the Annotation Ontology for image and document annotation, and the Gene Ontology. Each represents a specific discipline with a common terminology ("namespace") whose terminologies overlap in nanomedicine.

3) Form a core ontology to describe ENMs and share environmental, animal, and human nanotoxicology concepts among the different namespaces to construct a mapping of synonymous terms. The metadata should consistently describe data and models in a particular namespace.

4) For common terms having definitions that differ substantially, define the differences in underlying concepts by determining the additional concepts, attributes, or relationships that are embodied in different definitions to develop a mapping among the namespace terms, including the conceptual differences or nuances as modifiers. At this level, an ontology-user community has been defined that can begin to set user requirements for the larger nanoscience and nanotechnology community.

5) Expand the mapping among ontologies by including other ontologies, such as the Open Biological and Biomedical Ontology Foundry.

6) Develop mapping tools in conjunction with other mapping efforts. For example, a pilot project to develop an ontology crawler to enlarge and update mappings automatically may be undertaken with commercial, academic, and government participation.

7) Begin to develop a high-level ontology or taxonomy for the sciences that calls for participation by experts in major scientific disciplines that are integral to nanotechnology, such as physics, chemistry, material science, biology, and medicine.

8) Map the nanotechnology ontology to other scientific ontologies and develop standards for mapping based on the combined user-group experience.

9) Extend the mapping to metadata synonyms expressed in different natural languages, using existing or emerging ontologies for nanotechnology. Include mapping among terms having conceptual differences in terminology in different natural languages.

A Scenario for a Data-Sharing Infrastructure for Nanotechnology

The previous scenarios illustrated activities related to implementing required aspects of an informatics infrastructure: information content related to data quality and reliability, model quality and reliability, and the use of nomenclature and terminology for sharing, curating, and annotating information among disciplines. This scenario illustrates aspects of implementation of the informatics infrastructure itself and its capability to support collaboration.

1) Identify and assess existing pilot databases and knowledge bases that have been independently established to share data among specific sectors or by specific institutions.

2) Develop and adopt freely available software to federate databases to provide resources for the entire user community. Although a single, central database might conceivably satisfy all user requirements for sharing nanotechnology data across all stakeholder agencies and institutions, establishing a centralized, monolithic system is rarely possible, since each agency and entity must support its own particular mission and requirements. Such an approach would have difficulty in accommodating the heterogeneity of current database structures, security requirements, semantics and applications while respecting agency autonomy, and would have difficulty in providing uniform, expert data curation. An example of a possible federating system is the Cancer Biomedical Informatics Grid (caBIG®) sponsored by the NCI, supervised by the NCI Center for Bioinformatics and Information Technology (NCI-CBIIT), and used by caBIG's Nano Working Group. Others are NIH's National Center for Research Resources (NCRR) Biomedical Informatics Research Network (BIRN) and the EU-funded INFOBIOMED project on medical and biologic data interoperability and management.

3) Develop the infrastructure openly, leveraging available tools already in use where possible. At any stage, demonstrations or trials of an existing capability can be used to elicit needs and requirements from the relevant user community to inform the next iteration of updates and to improve software implementation. Examples of open scientific informatics software and tool development are Linked Open Data initiatives (W3C 2011b) which include measures of data quality and the OpenScience project (The OpenScience Project 2011).

4) Establish governance of the development process so that it is directly under the control of the stakeholder and user communities. Issues of data rights, intellectual property, and academic credit should be investigated early in the process to ensure satisfaction of user needs and requirements as part of the iterative cycle.

5) Develop, evaluate, and implement relevant technologies that enable the incorporation of new social and institutional mechanisms of interaction among users. One example is Google Wave, a digital application for rapid video development, annotation, and modification. Although Google has recently withdrawn Google Wave from the market, the tools for video annotation and update are still available for use (Google Wave 2010). Other tools, such as wikis, are now used in a number of scientific applications and are freely available through commercial vendors and by such applications as the CSN. Scientific web sites routinely allow feeds and blogs, remote participation in experiments at user facilities are becoming common, and sites for collaborative software development reflect standard practice. Cloud computing also has rapidly emerged as a new freely available tool for management of data archives and computational resources.

6) Where possible, prototyping activities should incorporate expert users throughout the data life cycle (from data generation to data-mining) to evaluate

possible new system capabilities, tools, or applications and to estimate resources required for the incorporation of those capabilities into the infrastructure.

REFERENCES

CSN. 2011. Nanocollaboratory. Collaboratory for Structural Nanobiology [online]. Available: http://nanobiology.utalca.cl or http://nanobiology.ncifcrf.gov [accessed Dec. 14, 2011].

Google Wave. 2010. The Google Wave Blog: News & Updates from the Google Wave Team [online]. Available: http://googlewave.blogspot.com/ [accessed July 8, 2011].

Haimes, Y.Y. 2009. P. xii in Risk Modeling, Assessment and Management, 3rd Ed. Hoboken, NJ: John Wiley & Sons.

IANH (International Alliance for NanoEHS Harmonization). 2011. International Alliance for NanoEHS Harmonization [online]. Available: http://www.nanoehsalliance.org/ [accessed Nov. 9, 2011].

InterNano. 2011. InterNano. Resources for Manufacturing. Nanoinformatics 2020 Roadmap [online]. Available: http://eprints.internano.org/607/ [accessed Nov. 9, 2011].

IOM (Institute of Medicine). 2011. Building a Framework for the Establishment of Regulatory Science for Drug Development, Y. Lebovitz, A.R. English, and A.B. Clairborne, eds. Washington, DC: National Academies Press.

JoVE. 2011. Journal of Visualized Experiments [online]. Available: http://www.jove.com/About.php?sectionid=0 [accessed July 8, 2011].

NanoHUB.org. 2011. NanoHUB [online]. Available: http://nanohub.org/[accessed Dec. 14, 2011].

Nanoinformatics. 2011a. Nanoinformatics: NanoinformaticsCommunity [online]. Available: http://www.internano.org/nanoinformatics/index.php/Nanoinformatics:NanoinformaticsCommunity [accessed Nov. 8, 2011].

Nanoinformatics. 2011b. Nanoinformatics 2011 Meeting Program, December 7-9, 2011, Arlington, VA [online]. Available: http://nanotechinformatics.org/program [accessed Nov. 8, 2011].

The OpenScience Project. 2011. The OpenScience Project [online]. Available: http://www.openscience.org/blog/?p=269 [accessed Nov. 8, 2011].

W3C (World Wild Web Consortium). 2011a. Semantic Web [online]. Available: http://www.w3.org/standards/semanticweb/ [accessed July 8, 2011].

W3C (World Wild Web Consortium). 2011b. LinkingOpenDate [online]. Available: http://www.w3.org/wiki/SweoIG/TaskForces/CommunityProjects/LinkingOpenData [accessed Nov. 8, 2011].

Wang, C.Y., W.E. Fu, H.L. Lin, and G.S. Peng. 2007. Preliminary study on nanoparticle sizes under the APEC technology cooperative framework. Mass Sci Technol. 18:487-495.

wwPDB. 2011. Worldwide wwPDB Protein Data Bank [online]. Available: http://www.wwpdb.org/[accessed Dec. 14, 2011].